Lecture Notes in Statistics 204

Edited by P. Bickel, P.J. Diggle, S. Fienberg, U. Gather,
I. Olkin, S. Zeger

For further volumes:
http://www.springer.com/series/694

To Rebecca Elisabeth and Sarah Johanna

Preface

Generalized estimating equations (GEE) were introduced by Liang and Zeger in a series of papers (see, e.g., Liang and Zeger, 1986; Zeger et al., 1985; Zeger and Liang, 1986) about 25 years ago. They have become increasingly popular in biometrical, econometrical, and psychometrical applications because they overcome the classical assumptions of statistics, i.e., independence and normality, which are too restrictive for many problems. The assumption of normality is, for example, violated if dichotomous data, e.g., positive or negative outcome, are considered, while the assumption of independence is not fulfilled in family studies or studies with repeated measurements. The development of more complex statistical methods like GEE is closely related to the progress in computer technology, because many modern regression approaches are based on iterative algorithms.

Originally, GEE have been proposed and further developed without considering them as special cases of quite general statistical methods. The main goal of this monograph therefore is to give a systematic presentation of the original GEE and some of its further developments. Subsequently, the emphasis is put on the unification of various GEE approaches. This is done by the use of two different estimation techniques, the pseudo maximum likelihood (PML) method and the generalized method of moments (GMM).

The PML approach was proposed by Gourieroux et al. in 1984b and further explained in additional work (Gourieroux and Monfort, 1993; Gourieroux et al., 1984a). The theory has been widely recognized by econometricians (see, e.g., Laroque and Salanie, 1989; Foncel et al., 2004) but to a lower extent by biostatisticians. A concise treatment of the PML theory has been given in Gourieroux and Monfort (1995a).

GMM was introduced by Hansen in 1982. It is very popular among econometricians because it provides a unified framework for the analysis of many well-known estimators, including least squares, instrumental variables (IV), maximum likelihood (ML), and PML. Several excellent book chapters and textbooks have been published (Hall, 1993; Ogaki, 1993), where readers may find many various introductory examples. As already indicated, the theory of GMM is very rich, see, e.g., the two special issues on GMM published in *J Bus Econ Stat* (1996, Vol. 14, Issue 3; 2002, Vol. 20, Issue 4), much richer than the part of the theory required for deriving GEE. For example, one important part of GMM theory is IV estimation (Baum et al., 2003; Stock et al., 2002), which is of no interest in GEE. As a result, for GEE only, "just identified" GMM models (Hall, 1993) are relevant. However, if only just identified models are of prime importance, other aspects of GMM theory, such as choice of the weight matrix, or 1-step, 2-step, or simultaneous GMM estimation do not play a role. For this short book, it was therefore difficult to decide whether GMM should be described comprehensively or whether the treatise should be concise and focus only on the aspects relevant for deriving GEE. The decision was to restrict the description of GMM to essential elements so that GEE remains the focus. For detailed descriptions of GMM, the reader may refer to the literature (Hall, 2005; Mátyás, 1999).

To increase readability, regularity conditions and technical details are not given, and many proofs are only sketched. Instead, references to the relevant literature discussing technical details are given for the interested reader. GEE have been proposed as methods for large samples. Therefore, only asymptotic properties will be considered throughout this book for both PML and GMM estimation.

The main aim of this monograph is the statistical foundation of the GEE approach using more general estimation techniques. This book could therefore be used as a basis for a course for graduate students in statistics, biostatistics, or econometrics. Knowledge of ML estimation is required at a level as usually imparted in undergraduate courses.

Organization of the Book

Several estimation techniques provide a quite general framework and include the GEE as a special case. An appealing approach is to embed the GEE into the framework of PML estimation (Gourieroux et al., 1984b; Gourieroux and Monfort, 1993). If the GEE are embedded into the PML approach, they can be interpreted as score equations derived from a specific likelihood model. The major advantage of this approach is that ML estimation is familiar to almost every statistician.

The PML approach is based on the exponential family. Chapts. 1 and 2 therefore deal with the linear and quadratic exponential family, respectively. The GEE method has been derived by Liang and Zeger (1986) as a generalization of the generalized linear model (GLM; McCullagh and Nelder, 1989), and Chapter 3 therefore deals with both univariate and multivariate GLM.

Because PML estimation can be considered a generalization of the ML approach, ML estimation is discussed in some detail in Chapt. 4. A crucial assumption of the ML method is the correct specification of the likelihood function. If it is misspecified, ML estimation may lead to invalid conclusions. The interpretation of ML estimators under misspecification and a test for detecting misspecifications are also considered in Chapt. 4.

Chapt. 5 deals with the PML method using the linear exponential family. It allows consistent estimation of the mean structure, even if the pre-specified covariance matrix is misspecified. Because the mean structure is consistently estimated, the approach is termed PML1. However, there is no free lunch, and a price has to be paid in terms of efficiency. Thus, a different covariance matrix, termed the robust covariance matrix, has to be used instead of the model-based covariance matrix, i.e., the Fisher information matrix. It can be shown that the robust covariance matrix always leads to an increased covariance matrix compared with the covariance matrix of the correctly specified model. Examples for the PML1 approach include the independence estimating equations (IEE) with covariance matrix equal to the identity matrix. Efficiency for estimating the mean structure may be improved if observations are weighted with fixed weights according to their degree of dependency. This approach results in the GEE for the mean structure (GEE1) with fixed covariance matrix.

Instead of using fixed weights, the weights might be estimated from the data. This results in increased power if the estimated covariance matrix is "closer" to the true covariance matrix than the pre-specified covariance matrix. This idea leads to the quasi generalized PML (QGPML) approach, which will be discussed in Chapt. 6. An important aspect is that under suitable regularity conditions, no adjustments have to be made for the extra variability, that is introduced by estimating the possibly misspecified covariance matrix. Even more, the QGPML estimator is efficient for the mean structure in the sense of Rao-Cramér if both the mean structure and the association structure, e.g., the covariance structure, are correctly specified. The likelihood function might thus be misspecified. Examples of QGPML estimation include the IEE with estimated variances, the GEE1 with estimated working covariance matrix, and the well-known GEE1 with estimated working correlation matrix. Examples for common weight matrices, i.e., working covariance and working correlation structures, are discussed as well as time dependent parameters and models for ordinal dependent variables.

Chapt. 7 deals with the consistent estimation of both the mean and the association structure using the PML approach. It is based on the quadratic exponential family and therefore termed the PML2 method. Special cases include the GEE2 for the mean and the correlation coefficient as the interpretable measure of association and, for dichotomous dependent variables, the GEE2 for the mean and the log odds ratio as the interpretable measure of association. In the first special case, the estimating equations are formulated in the second centered moments, while the second ordinary moments are used as the measure of association in the second special case.

The two GEE2 approaches considered in Chapt. 7 require the simultaneous solution of the estimating equations for the mean structure and the association structure. This has three disadvantages. First, the computational effort is substantially larger when both estimating equations are solved jointly. Second, if the association structure is misspecified, the parameter estimates of the mean structure are still estimated consistently if the estimating equations are solved separately, i.e., in a two-stage approach. The simultaneous estimation of the estimating equations for the mean and the association structure may lead to biased parameter estimates if the mean structure is correctly specified but the association structure is misspecified. Therefore, one aim is to separate the estimating equations into a two-step approach. Third, GEE2 using the second standardized moments, i.e., the correlation coefficient as the measure of association, cannot be derived using the PML2 method.

All three disadvantages can be overcome by estimating equations that can be derived using GMM (Chapt. 8). Specifically, GMM allow the formulation of GEE2 using the correlation as the measure of association in two separate estimating equations. Similarly, the alternating logistic regression (ALR) that uses the log odds ratio through the ordinary second moments as the measure of association can be formulated as a special case of GMM. Therefore, GMM will be considered in the last chapter. The use of GMM is illustrated with the linear regression model as the introductory example. Second, the IEE are derived within the GMM framework. Third, the GEE2 using the correlation as the measure of association is considered. Finally, the ALR are derived as a special case of GMM. Again, we stress that only a small portion of the rich theory of GMM is required for deriving the GEE2 models, and we restrict the discussion of GMM to the needs in this monograph. Specifically, we do not consider IV estimation, and we focus on "just identified" models.

Acknowledgments

First and foremost, I have to thank my doctoral adviser Gerhard Arminger, who got me interested in generalized estimating equations (GEE) almost 20 years ago. He exposed me to the econometrics literature and taught me how to bridge the gap between biostatistics and econometrics. The next substantial impetus was provided by Ludwig Fahrmeir, who asked the important question why no price has to be paid in estimating the covariance matrix in GEE even if a working correlation matrix is estimated. I did not have a convincing argument at hand, and it was exactly this question to which I dedicated a lot of time.

While on a sabbatical in Cleveland, Ohio, in 2005, Robert C. Elston strongly encouraged me to publish a short monograph on GEE with all related results on pseudo maximum likelihood estimation. In 2010, I took my next sabbatical and during this time I have been able to complete the work presented here. My perfect host at the Institute Montefiore was Kristel Van Steen, and she deserves special thanks.

I am also much obliged to John Kimmel and Marc Strauss, former and current senior editors of Statistics at Springer.

Finally, acknowledgments go to my family. My wife has always smiled when I have tried to explain "some great formula" to her. My daughters have always motivated me by stating that the text and the formulae look quite impressive. I am sure they will also make me smile in the future.

Lübeck, April 2011 *Andreas Ziegler*

Contents

Chapter 1
The linear exponential family

Several estimating equations belonging to the class of generalized estimating equations for the mean structure, termed GEE1, can be derived as special cases of the pseudo maximum likelihood 1 (PML1) method. PML1 estimation is based on the linear exponential family, and this class of distributions is therefore discussed in this chapter. In Sect. 1.1, the linear exponential family is defined in the canonical form with a natural parameter. Moments of the exponential family can be easily obtained by differentiation (Sect. 1.2), which can in turn be used for parameterizing the exponential family in the mean structure (Sect. 1.3). Some properties of the linear exponential family are required for PML1 estimation in Chapt. 5, and they are presented in Sect. 1.4. In Sects. 1.5 and 1.6, several examples for univariate and multivariate distributions belonging to the linear exponential family are given to illustrate the broad applicability of the linear exponential family. Finally, the relationship to the parameterization in generalized linear models is established in Sect. 1.7.

1.1 Definition

Some general notation is given at the very beginning. All vectors are considered to be column vectors. Vectors and matrices are given in bold, and $'$ indicates transposition of vectors and matrices. For count indices, $'$ is also used. It can be distinguished from the transposition easily because it stands with a non-bold letter. True parameters of an underlying distribution will be denoted by "||" throughout.

We start off with a T-dimensional random vector $\boldsymbol{y}$. $\boldsymbol{y}$ will be directly connected with a parameter $\boldsymbol{\vartheta}$ of the same size via $\boldsymbol{\vartheta}'\boldsymbol{y}$, leading to the canonical or natural form of the linear exponential family. The distribution may be parameterized in an additional matrix of fixed nuisance parameters $\boldsymbol{\Psi}$.

Definition 1.1 (Simple linear exponential family). Let $\boldsymbol{y} \in \mathbb{R}^T$ be a random vector, $\boldsymbol{\vartheta} \in \Theta \subset \mathbb{R}^T$ be the parameter vector of interest, $\boldsymbol{\Psi} \in \mathbb{R}^{T \times T}$ be a positive definite matrix of fixed nuisance parameters, $b: \mathbb{R}^T \times \mathbb{R}^{T \times T} \to \mathbb{R}$, and $d: \mathbb{R}^T \times \mathbb{R}^{T \times T} \to \mathbb{R}$ some functions. A T-dimensional distribution belongs to the T-dimensional simple linear exponential family, if its density (meant to include probability mass functions for discrete data) is given by

$$f(\boldsymbol{y}||\boldsymbol{\vartheta}, \boldsymbol{\Psi}) = \exp\Big(\boldsymbol{\vartheta}'\boldsymbol{y} + b(\boldsymbol{y}, \boldsymbol{\Psi}) - d(\boldsymbol{\vartheta}, \boldsymbol{\Psi})\Big) . \tag{1.1}$$

$\boldsymbol{\vartheta}$ is termed the natural parameter, and Θ is the natural parameter space.

Therefore, Θ is the set of all $\boldsymbol{\vartheta} \in \mathbb{R}^T$ for which

$$0 < \exp\{d(\boldsymbol{\vartheta}, \boldsymbol{\Psi})\} = \int_{\mathbb{R}^T} \exp\{\boldsymbol{\vartheta}'\boldsymbol{y} + b(\boldsymbol{y}, \boldsymbol{\Psi})\} \, d\boldsymbol{y} < \infty \tag{1.2}$$

holds. $d(\boldsymbol{\vartheta}, \boldsymbol{\Psi})$ can be considered a normalization constant. In Theorem 1.2, it will be shown that $d(\boldsymbol{\vartheta}, \boldsymbol{\Psi})$ is the cumulant generating function of $f(\boldsymbol{y}||\boldsymbol{\vartheta}, \boldsymbol{\Psi})$, i.e., the logarithm of the moment-generating function (see, e.g., Lehmann and Casella, 1998, p. 28, Theorem 5.10).

1.2 Moments

Theorem 1.2. *The random vector $\boldsymbol{y}$ is assumed to have a density belonging to the T-dimensional simple linear exponential family.*

1. *If $\phi(\boldsymbol{y})$ is an integrable function with values in $\mathbb{R}$, then all higher order derivatives of*

$$\int_{\mathbb{R}^T} \phi(\boldsymbol{y}) \exp\{\boldsymbol{\vartheta}'\boldsymbol{y} + b(\boldsymbol{y}, \boldsymbol{\Psi})\} d\boldsymbol{y} \tag{1.3}$$

with respect to $\boldsymbol{\vartheta}$ exist, and differentiation and integration can be exchanged.

2.

$$\mathbb{E}(\boldsymbol{y}) = \boldsymbol{\mu} = \frac{\partial d(\boldsymbol{\vartheta}, \boldsymbol{\Psi})}{\partial \boldsymbol{\vartheta}} , \tag{1.4}$$

$$\mathbb{V}ar(\boldsymbol{y}) = \boldsymbol{\Sigma} = \frac{\partial^2 d(\boldsymbol{\vartheta}, \boldsymbol{\Psi})}{\partial \boldsymbol{\vartheta} \partial \boldsymbol{\vartheta}'} . \tag{1.5}$$

3. *If $\boldsymbol{\Sigma} > \mathbf{0}$, then $\ln f(\boldsymbol{y}||\boldsymbol{\vartheta}, \boldsymbol{\Psi})$ is strongly concave in $\boldsymbol{\vartheta}$, i.e., $\ln f$ is concave.*

The domain of $\boldsymbol{\mu}$ is denoted by $\Delta, \Delta \subset \mathbb{R}^T$.

Proof. 1. and 2. are standard results (see, e.g., Lehmann and Casella, 1998, p. 27, Theorem 5.8).
We only sketch the proof of 3.: $\ln f(\boldsymbol{y}||\boldsymbol{\vartheta}, \boldsymbol{\Psi})$ is strongly concave because

$$-\frac{\partial^2 \ln f(\boldsymbol{y}||\boldsymbol{\vartheta}, \boldsymbol{\Psi})}{\partial\boldsymbol{\vartheta}\partial\boldsymbol{\vartheta}'} = \frac{\partial^2 d(\boldsymbol{\vartheta}, \boldsymbol{\Psi})}{\partial\boldsymbol{\vartheta}\partial\boldsymbol{\vartheta}'} = \boldsymbol{\Sigma} = \mathbb{V}\mathrm{ar}(\boldsymbol{y}) > \mathbf{0}\,.$$

□

1.3 Parameterization in the mean

Throughout the following, $\boldsymbol{\Sigma} > \mathbf{0}$ is assumed, i.e., distributions are not degenerated. In PML1 estimation (Chapt. 5), the mean structure parameterization is used instead of the parameterization in the natural parameter vector. Because $\boldsymbol{\Sigma}$ is positive definite, and $\boldsymbol{\Psi}$ is a matrix of fixed nuisance parameters, the corresponding function to $\boldsymbol{\mu} = \partial d(\boldsymbol{\vartheta}, \boldsymbol{\Psi})/\partial\boldsymbol{\vartheta}$ from Θ to Δ is bijective. Therefore, a one-to-one inverse mapping $\boldsymbol{c}$ from Δ to Θ exists such that

$$\boldsymbol{c}^{-1}(\boldsymbol{\vartheta}, \boldsymbol{\Psi}) = \frac{\partial d(\boldsymbol{\vartheta}, \boldsymbol{\Psi})}{\partial\boldsymbol{\vartheta}} = \boldsymbol{\mu}, \qquad \boldsymbol{\vartheta} = \boldsymbol{c}(\boldsymbol{\mu}, \boldsymbol{\Psi})\,. \tag{1.6}$$

Equation 1.1 can subsequently be rewritten as

$$\tilde{f}(\boldsymbol{y}||\boldsymbol{\mu}, \boldsymbol{\Psi}) = \exp\left\{\boldsymbol{c}(\boldsymbol{\mu}, \boldsymbol{\Psi})'\boldsymbol{y} + a(\boldsymbol{\mu}, \boldsymbol{\Psi}) + b(\boldsymbol{y}, \boldsymbol{\Psi})\right\}\,, \tag{1.7}$$

where $a(\boldsymbol{\mu}, \boldsymbol{\Psi}) = -d\big(\boldsymbol{c}(\boldsymbol{\mu}, \boldsymbol{\Psi}), \boldsymbol{\Psi}\big)$. Examples for the functions $\boldsymbol{c}$, a, and b are given in Sects. 1.5 and 1.6.

1.4 Selected properties

In the following sections and chapters, several properties of the linear exponential family are required, and they are derived in this section. The properties are formulated first, and their proofs are given at the end of this section. For a different formulation of the properties, the reader may refer to Gourieroux et al. (1984b).

Property 1.3.

$$\boldsymbol{\Sigma}^{-1} = \frac{\partial \boldsymbol{c}(\boldsymbol{\mu}, \boldsymbol{\Psi})'}{\partial\boldsymbol{\mu}} = \frac{\partial \boldsymbol{c}(\boldsymbol{\mu}, \boldsymbol{\Psi})}{\partial\boldsymbol{\mu}'} = \boldsymbol{h}(\boldsymbol{\mu}, \boldsymbol{\Psi})\,. \tag{1.8}$$

$\boldsymbol{h}$ is termed the variance function.

Property 1.4.

$$\left(\frac{\partial a(\boldsymbol{\mu},\boldsymbol{\Psi})}{\partial \boldsymbol{\mu}} + \frac{\partial \boldsymbol{c}(\boldsymbol{\mu},\boldsymbol{\Psi})'}{\partial \boldsymbol{\mu}}\boldsymbol{y}\right) = \left(-\frac{\partial \boldsymbol{c}(\boldsymbol{\mu},\boldsymbol{\Psi})'}{\partial \boldsymbol{\mu}}\boldsymbol{\mu} + \frac{\partial \boldsymbol{c}(\boldsymbol{\mu},\boldsymbol{\Psi})'}{\partial \boldsymbol{\mu}}\boldsymbol{y}\right). \tag{1.9}$$

Property 1.5.

$$\frac{\partial^2 a(\boldsymbol{\mu},\boldsymbol{\Psi})}{\partial \boldsymbol{\mu}\partial \boldsymbol{\mu}'} + \sum_{l=1}^{T} \frac{\partial^2 c_l(\boldsymbol{\mu},\boldsymbol{\Psi})}{\partial \boldsymbol{\mu}\partial \boldsymbol{\mu}'} y_l = \sum_{l=1}^{T} \frac{\partial^2 c_l(\boldsymbol{\mu},\boldsymbol{\Psi})}{\partial \boldsymbol{\mu}\partial \boldsymbol{\mu}'}(y_l - \mu_l) - \frac{\partial \boldsymbol{c}'(\boldsymbol{\mu},\boldsymbol{\Psi})}{\partial \boldsymbol{\mu}}, \tag{1.10}$$

where $y_l, c_l(\boldsymbol{\mu},\boldsymbol{\Psi})$, and μ_l are the lth component of $\boldsymbol{y}, \boldsymbol{c}(\boldsymbol{\mu},\boldsymbol{\Psi})$, and $\boldsymbol{\mu}$, respectively.

Property 1.6. For any $\boldsymbol{\mu} \in \Delta$ and for fixed $\boldsymbol{\mu}_0 \in \Delta$, the following is true:

$$a(\boldsymbol{\mu},\boldsymbol{\Psi}) + \boldsymbol{c}(\boldsymbol{\mu},\boldsymbol{\Psi})'\boldsymbol{\mu}_0 \leq a(\boldsymbol{\mu}_0,\boldsymbol{\Psi}) + \boldsymbol{c}(\boldsymbol{\mu}_0,\boldsymbol{\Psi})'\boldsymbol{\mu}_0\,. \tag{1.11}$$

The equality holds if and only if $\boldsymbol{\mu} = \boldsymbol{\mu}_0$.

Proof (Property 1.3). Property 1.3 directly follows from Eq. 1.5 using Eq. 1.6. □

Proof (Property 1.4). By using the chain rule, we obtain from Eq. (1.4):

$$\boldsymbol{\mu} = \frac{\partial d(\boldsymbol{\vartheta},\boldsymbol{\Psi})}{\partial \boldsymbol{c}(\boldsymbol{\mu},\boldsymbol{\Psi})} = -\frac{\partial \boldsymbol{\mu}'}{\partial \boldsymbol{c}(\boldsymbol{\mu},\boldsymbol{\Psi})} \cdot \frac{\partial a(\boldsymbol{\mu},\boldsymbol{\Psi})}{\partial \boldsymbol{\mu}}\,. \tag{1.12}$$

With $\boldsymbol{\Sigma} > \boldsymbol{0}$, property 1.4 follows from Eq. 1.12 by adding $\dfrac{\partial \boldsymbol{c}(\boldsymbol{\mu},\boldsymbol{\Psi})'}{\partial \boldsymbol{\mu}}\boldsymbol{y}$ because of Eq. 1.8. □

Proof (Property 1.5). In the first step, we obtain

$$\frac{\partial \boldsymbol{c}(\boldsymbol{\mu},\boldsymbol{\Psi})'}{\partial \boldsymbol{\mu}}\boldsymbol{\mu} + \frac{\partial a(\boldsymbol{\mu},\boldsymbol{\Psi})}{\partial \boldsymbol{\mu}} = \boldsymbol{0} \tag{1.13}$$

from Eqs. 1.8 and 1.12. In the second step, the derivative of Eq. 1.13 is taken with respect to $\boldsymbol{\mu}'$. Property 1.5 now follows in the third step by adding $\sum_{l=1}^{T}\frac{\partial^2 c_l(\boldsymbol{\mu},\boldsymbol{\Psi})}{\partial \boldsymbol{\mu}\partial \boldsymbol{\mu}'} y_l$ on both sides of the resulting equation. □

Proof (Property 1.6). The proof uses the Kullback-Leibler information criterion, which will be discussed in some detail in Chapt. 4. The Kullback-Leibler information is a distance and therefore non-negative. Subsequently, Kullback's inequality (see, e.g., Rao, 1973, p. 59, ii) for arbitrary densities $f(y)$ and $g(y)$ is given by

$$\int \left(\ln \frac{f(y)}{g(y)}\right) f(y)dy \geq 0\,.$$

Equality holds almost surely (a.s.) if and only if $f(y) = g(y)$. Let $f(y) = \tilde{f}(\boldsymbol{y}||\boldsymbol{\mu}_0, \boldsymbol{\Psi})$ and $g(y) = \tilde{f}(\boldsymbol{y}||\boldsymbol{\mu}, \boldsymbol{\Psi})$. It follows

$$\int \left(\ln \frac{\tilde{f}(\boldsymbol{y}||\boldsymbol{\mu}_0, \boldsymbol{\Psi})}{\tilde{f}(\boldsymbol{y}||\boldsymbol{\mu}, \boldsymbol{\Psi})} \right) \tilde{f}(\boldsymbol{y}||\boldsymbol{\mu}_0, \boldsymbol{\Psi}) \boldsymbol{dy} \geq 0 \,.$$

After cancelling $b(\boldsymbol{y}, \boldsymbol{\Psi})$ and solving the logarithm, we obtain

$$\int \left(\boldsymbol{c}(\boldsymbol{\mu}_0, \boldsymbol{\Psi})'\boldsymbol{y} + a(\boldsymbol{\mu}_0, \boldsymbol{\Psi}) \right) \tilde{f}(\boldsymbol{y}||\boldsymbol{\mu}_0, \boldsymbol{\Psi}) \boldsymbol{dy} \geq \int \left(\boldsymbol{c}(\boldsymbol{\mu}, \boldsymbol{\Psi})'\boldsymbol{y} + a(\boldsymbol{\mu}, \boldsymbol{\Psi}) \right) \tilde{f}(\boldsymbol{y}||\boldsymbol{\mu}_0, \boldsymbol{\Psi}) \boldsymbol{dy} \,.$$

Property 1.6 follows from solving the integral and by noting that $\tilde{f}(\boldsymbol{y}||\boldsymbol{\mu}_0, \boldsymbol{\Psi})$ is a density with expectation $\boldsymbol{\mu}_0$. □

1.5 Examples for univariate linear exponential families

In this section, we give several examples for the univariate linear exponential family. The Poisson, the binomial, and the negative binomial distributions are discrete distributions, while the univariate normal and the gamma distribution are continuous. Other examples for the linear exponential family are the beta, the Dirichlet, and the geometric distribution. The Weibull distributions and the Cauchy distributions do not belong to the class of the linear exponential family.

In the examples, three different notations are used to characterize different parameterizations. $\approx$ denotes the standard parameterization of a distribution, i.e., the parameterization that is usually taught in basic statistics courses. In contrast, $\tilde{}$ stands for the mean parameterization, and, finally, no tilde is used for the parameterization in the natural parameter.

Example 1.7 (Poisson distribution). The probability function of the discrete Poisson distribution, denoted by $Po(\lambda)$, is given by

$$\overset{\approx}{f}(y||\lambda) = \lambda^y e^{-\lambda}/y! = \exp\left((\ln\lambda)\, y - \ln y! - \lambda\right) \qquad (1.14)$$

for $\lambda > 0$ and $y \in \mathbb{N}_0$. The density is independent of Ψ, thus $\Psi = 1$. The natural parameter is $\vartheta = \ln\lambda$. The simple form of the linear exponential family is therefore given by

$$f(y||\vartheta) = \overset{\approx}{f}(y||\lambda \equiv e^{\vartheta}) = \exp\left(\vartheta\, y - \ln y! - e^{\vartheta}\right)$$

with $b(y) = -\ln y!$ and $d(\vartheta) = e^{\vartheta}$. Using Eqs. 1.4 and 1.5, we obtain $\mu = \frac{\partial d(\vartheta)}{\partial \vartheta} = e^{\vartheta} = \lambda$, $\Delta = \mathbb{R}_+$ and $\mathbb{V}\mathrm{ar}(y) = \frac{\partial^2 d(\vartheta)}{\partial \vartheta^2} = e^{\vartheta} = \lambda$. The mean parameterization of the Poisson distribution is therefore given by

$$\tilde{f}(y||\mu) = \exp\left((\ln\mu)\, y - \mu - \ln y!\right)$$

with $c(\mu) = \ln\mu$ and $a(\mu) = -\mu$.

Example 1.8 (Binomial distribution). The probability function of the discrete binomial distribution, denoted by $B(n,p)$, for fixed n and $p \in]0,1[$ is given by

$$\overset{\approx}{f}(y||n,p) = \binom{n}{y} p^y (1-p)^{n-y} = \exp\Big(\mathrm{logit}(p)\cdot y + \ln\binom{n}{y} + n\ln(1-p)\Big) \quad (1.15)$$

for $y = 0,\ldots,n$ and $\mathrm{logit}(x) = \ln\frac{x}{1-x}$. The natural parameter is $\vartheta = \mathrm{logit}(p)$, and one obtains

$$f(y||\vartheta) = \exp\left(\vartheta y + \ln\binom{n}{y} - n\ln(1+e^{\vartheta})\right)$$

with $\Psi = 1$, $b(y) = \ln\binom{n}{y}$, and $d(\vartheta) = -n\ln(1+e^{\vartheta})$ because of $p = e^{\vartheta}/(1+e^{\vartheta})$ and $1-p = 1/(1+e^{\vartheta})$. Note that $\Psi = 1$.

Furthermore, we obtain $\mu = \mathbb{E}(y) = n\,e^{\vartheta}/\left(1+e^{\vartheta}\right) = np$ and $\mathbb{V}\mathrm{ar}(y) = ne^{\vartheta}/\left(1+e^{\vartheta}\right)^2 = np\,(1-p)$ by differentiating d with respect to θ. In addition, $\Delta = [0,n]$. Finally, the mean parameterization of the binomial distribution is given by

$$\tilde{f}(y||\mu) = \exp\left(\ln\left(\tfrac{\mu}{n-\mu}\right)y + n\ln\left(\tfrac{n-\mu}{n}\right) + \ln\binom{n}{y}\right)$$

with $c(\mu) = \ln\left(\frac{\mu}{n-\mu}\right)$, $a(\mu) = n\ln\left(\frac{n-\mu}{n}\right)$, and $b(y) = \ln\binom{n}{y}$.

Example 1.9 (Negative binomial distribution – Pascal distribution). The probability function of the (discrete) $NB(\Psi,p)$ distribution with $0 < p < 1$ is given by

$$\overset{\approx}{f}(y|p,\Psi) = \binom{\Psi+y-1}{y} p^{\Psi}(1-p)^y$$

for y and $\Psi \in \mathbb{N}_0$. The natural parameter is $\vartheta = \ln(1-p)$, thus $p = 1-e^{\vartheta}$, and the parameterization in ϑ is

$$f(y||\vartheta) = \exp\left(\vartheta y + \ln\binom{\Psi+y-1}{y} + \Psi\ln(1-e^{\vartheta})\right)$$

with $b(y,\Psi) = \ln\binom{\Psi+y-1}{y}$ and $d(\vartheta,\Psi) = -\Psi\ln(1-e^{\vartheta})$. Correspondingly, the parameterization in the mean $\mu = \mathbb{E}(y) = \Psi\frac{1-p}{p}$ is given by

$$\tilde{f}(y||\mu,\Psi) = \exp\left(c(\mu,\Psi)y + \Psi\ln(\tfrac{\Psi}{\mu+\Psi}) + \ln\binom{\Psi+y-1}{y}\right)$$

with $c(\mu,\Psi) = \ln\left(\frac{\mu}{\Psi+\mu}\right)$, $a(\mu,\Psi) = \Psi\ln(\frac{\Psi}{\mu+\Psi})$, and $b(y,\Psi) = \ln\binom{\Psi+y-1}{y}$. Finally, $\mathbb{V}\mathrm{ar}(y) = \Psi\frac{1-p}{p^2} = \frac{\mu}{p}$.

Example 1.10 (Univariate normal distribution). The density of the univariate normal distribution $N(\mu,\Psi)$ is given by

$$\tilde{\tilde{f}}(y||\mu,\Psi) = \tilde{f}(y||\mu,\Psi) = \frac{1}{\sqrt{2\pi\Psi}}\exp\left(-\frac{1}{2}\frac{(y-\mu)^2}{\Psi}\right) \tag{1.16}$$

for $y \in \mathbb{R}$, $\mu \in \Delta \subset \mathbb{R}$, and $\Psi \in \mathbb{R}_+$. The parameterization of Eq. 1.16 is thus identical to the parameterization in the mean. Here, $a(\mu,\Psi) = -\mu^2/(2\Psi)$, $b(y,\Psi) = -\frac{1}{2}\ln(2\pi\Psi) - y^2/(2\Psi)$, and $d(\vartheta,\Psi) = \vartheta^2\Psi/2$. We obtain $\mathbb{E}(y) = \mu$ and $\mathbb{V}\mathrm{ar}(y) = \Psi$. The natural parameter is $\vartheta = \mu/\Psi = c(\mu,\Psi)$, and the density in the natural parameter is given by

$$f(y||\vartheta,\Psi) = \exp\left(\theta y - \frac{1}{2}\ln(2\pi\Psi) - y^2/(2\Psi) - \vartheta^2\Psi/2\right).$$

Example 1.11 (Gamma distribution). The density of the $G(\alpha,\Psi)$ distribution is given by

$$\tilde{\tilde{f}}(y|\alpha,\Psi) = \alpha^\Psi y^{\Psi-1}e^{-\alpha y}/\Gamma(\Psi)$$

for $y \in \mathbb{R}_+$, $\alpha \in \mathbb{R}_+$, and $\Psi \in \mathbb{R}_+$, where $\Gamma(\Psi)$ denotes the Gamma function. The natural parameter is $\vartheta = -\alpha$. Mean and variance are given by $\mu = \mathbb{E}(y) = \Psi/\alpha = -\Psi/\vartheta$, and $\mathbb{V}\mathrm{ar}(y) = \Psi/\alpha^2$, respectively. As a result, the parameterizations in the natural parameter and in the mean are given by

$$f(y||\theta,\Psi) = \exp\Big(\theta y + (\Psi-1)\ln y - \ln\Gamma(\Psi) + \Psi\ln(-\vartheta)\Big), \qquad \text{and}$$
$$f(y||\mu,\Psi) = \exp\Big(-\tfrac{\Psi}{\mu}\,y + \Psi\ln\left(\tfrac{\Psi}{\mu}\right) + (\Psi-1)\ln y - \ln\Gamma(\Psi)\Big),$$

respectively, with $b(y,\Psi) = (\Psi-1)\ln y - \ln\Gamma(\Psi)$, $d(\vartheta,\Psi) = -\Psi\ln(-\vartheta)$, $c(\mu,\Psi) = -\Psi/\mu$, and $a(\mu,\Psi) = \Psi\ln\left(\Psi/\mu\right)$.

The Gamma distribution belongs to the two-parameter linear exponential family, and therefore $\Psi \neq 1$, in general. For $\Psi = 1$, one obtains the exponential distribution.

The nuisance parameter Ψ equals 1 for the Poisson and the binomial distribution. In the other examples, Ψ is a differentiable function and functionally dependent on μ and Σ. For the normal distribution, the variance depends on the nuisance parameter Ψ but not on the mean μ. Finally, for the Gamma distribution, Ψ is the inverse of the squared coefficient of variation.

1.6 Examples for multivariate linear exponential families

In this section, we consider the two most popular multivariate linear exponential families, i.e., the continuous multivariate normal distribution and the discrete multinomial distribution.

Example 1.12 (Multivariate normal distribution). The density of the multivariate normal distribution $N_T(\boldsymbol{\mu}, \boldsymbol{\Psi})$ is given by

$$\tilde{f}(\boldsymbol{y}||\boldsymbol{\mu}, \boldsymbol{\Psi}) = \tilde{\tilde{f}}(\boldsymbol{y}||\boldsymbol{\mu}, \boldsymbol{\Psi}) = (2\pi)^{-\frac{T}{2}} \det(\boldsymbol{\Psi})^{-\frac{1}{2}} \exp\Big(-\tfrac{1}{2}(\boldsymbol{y}-\boldsymbol{\mu})'\boldsymbol{\Psi}^{-1}(\boldsymbol{y}-\boldsymbol{\mu})\Big) \tag{1.17}$$

for $\boldsymbol{y} \in \mathbb{R}^T, \boldsymbol{\mu} \in \Delta \subset \mathbb{R}^T$, and a positive definite $T \times T$ matrix $\boldsymbol{\Psi}$. The natural parameter is $\boldsymbol{\vartheta} = \boldsymbol{c}(\boldsymbol{\mu}, \boldsymbol{\Psi}) = \boldsymbol{\Psi}^{-1}\boldsymbol{\mu}$. $d(\boldsymbol{\vartheta}, \boldsymbol{\Psi}) = \frac{1}{2}\boldsymbol{\vartheta}'\boldsymbol{\Psi}\boldsymbol{\vartheta}$ can be used for determining the first two moments, which are given by $\boldsymbol{\mu} = \mathbb{E}(\boldsymbol{y})$ and $\boldsymbol{\Sigma} = \mathbb{V}\mathrm{ar}(\boldsymbol{y}) = \boldsymbol{\Psi}$. Furthermore, $b(\boldsymbol{y}, \boldsymbol{\Psi}) = -\frac{T}{2}\ln(2\pi) - \frac{1}{2}\boldsymbol{y}'\boldsymbol{\Psi}^{-1}\boldsymbol{y} - \frac{1}{2}\ln(\det(\boldsymbol{\Psi}))$, and $a(\boldsymbol{\mu}, \boldsymbol{\Psi}) = -\frac{1}{2}\boldsymbol{\mu}'\boldsymbol{\Psi}^{-1}\boldsymbol{\mu}$. Subsequently, the parameterization in the natural parameter $\boldsymbol{\vartheta}$ can be shown to be

$$f(\boldsymbol{y}||\boldsymbol{\vartheta}, \boldsymbol{\Psi}) = \exp\Big(\boldsymbol{\vartheta}'\boldsymbol{y} - \tfrac{T}{2}\ln(2\pi) - \tfrac{1}{2}\boldsymbol{y}'\boldsymbol{\Psi}^{-1}\boldsymbol{y} - \tfrac{1}{2}\ln(\det(\boldsymbol{\Psi})) - \tfrac{1}{2}\boldsymbol{\vartheta}'\boldsymbol{\Psi}\boldsymbol{\vartheta}\Big).$$

Example 1.13 (Multinomial distribution). We consider the multinomial distribution $Mu_T(n, \boldsymbol{\pi})$ for $\boldsymbol{y} = (y_1, \ldots, y_{T+1})'$, $y_t \in \mathbb{N}_0$ and $\sum_{t=1}^{T+1} y_t = n$. The parameters $\boldsymbol{\pi} = (\pi_1, \ldots, \pi_T)'$, $\pi_{T+1} = 1 - \sum_{t=1}^{T} \pi_t$, are interpreted as probabilities in analogy to the binomial distribution. Note that y_{T+1} can be neglected as long as n is fixed.

The probability function of the $Mu_T(n, \boldsymbol{\pi})$ distribution is given by

$$\tilde{\tilde{f}}(\boldsymbol{y}||\boldsymbol{\pi}) = C(\boldsymbol{y}) \prod_{t=1}^{T} (\pi_t^{y_t}) \Big(1 - \sum\nolimits_{t=1}^{T} \pi_t\Big)^{n - \sum_{t=1}^{T} y_t},$$

where $C(\boldsymbol{y}) = n! / \prod_{t=1}^{T+1} y_t!$ is the coefficient of the multinomial distribution. The density is independent of $\boldsymbol{\Psi}$, thus $\boldsymbol{\Psi} = \boldsymbol{I}_T$, where $\boldsymbol{I}_T$ denotes the T-dimensional identity matrix.

The natural parameter is given by

$$\vartheta_t = \ln\Big(\pi_t / (1 - \sum\nolimits_{t'=1}^{T} \pi_{t'})\Big)$$

for $t = 1, \ldots, T$. This is equivalent to

$$\pi_t = \frac{e^{\vartheta_t}}{1 + \sum_{t'=1}^{T} e^{\vartheta_{t'}}} \qquad \text{and} \qquad \pi_{T+1} = 1 - \sum_{t=1}^{T} \pi_t = \frac{1}{1 + \sum_{t=1}^{T} e^{\vartheta_t}}.$$

For $t \neq t'$, mean, variance, and covariance are given by

$$\mu_t = \mathbb{E}(y_t) = n\pi_t, \quad \mathbb{V}\text{ar}(y_t) = n\pi_t(1-\pi_t), \quad \mathbb{C}\text{ov}\,(y_t, y_{t'}) = -n\pi_t\pi_{t'}\,.$$

With $b(\boldsymbol{y}) = \ln C(\boldsymbol{y})$ and $d(\boldsymbol{\vartheta}) = n\ln(1+\sum_{l=1}^{T} e^{\vartheta_l}) \equiv -n\ln(1-\sum_{l=1}^{T}\pi_l)$, the parameterization in the natural parameter $\boldsymbol{\vartheta}$ is given by

$$f(\boldsymbol{y}||\boldsymbol{\vartheta}) = \exp\Big(\boldsymbol{\vartheta}'\boldsymbol{y} + \ln C(\boldsymbol{y}) + n\ln(1-\sum\nolimits_{l=1}^{T}\pi_l)\Big)\,.$$

Finally, we obtain the parameterization in the mean vector $\boldsymbol{\mu}$ by using the functions $a(\boldsymbol{\mu}) = 0$ and $\boldsymbol{c}(\boldsymbol{\mu}) = \big(\ln\frac{\mu_1}{n}, \ldots, \ln\frac{\mu_T}{n}\big)'$ as

$$\tilde{f}(\boldsymbol{y}||\boldsymbol{\mu}) = \exp\Big(\big(\ln\tfrac{\mu_1}{n}, \ldots, \ln\tfrac{\mu_T}{n}\big)'\boldsymbol{y} + \ln C(\boldsymbol{y})\Big)\,.$$

1.7 Relationship to the parameterization in univariate generalized linear models

In their excellent textbook on generalized linear models (GLM), McCullagh and Nelder (1989) used a definition for the univariate linear exponential family that slightly differs from Eq. 1.1 of definition 1.1. Specifically, they gave the density as

$$f(y||\tilde{\vartheta}, \Psi) = \exp\Big(\big(\tilde{\vartheta}y - b^\star(\tilde{\vartheta})\big)/a^\star(\Psi) + c^\star(y, \Psi)\Big)\,. \tag{1.18}$$

Here, the parameter $\tilde{\vartheta}$ is proportional to the natural parameter ϑ, and Ψ is, as before, the nuisance parameter. The symbols $a^\star, b^\star$, and $c^\star$ were chosen both to resemble the notation of McCullagh and Nelder (1989) and to avoid double notation with the previous sections.

In most applications, $a^\star(\Psi)$ is simplified to $a^\star(\Psi) = a^\star \cdot \Psi$ with known weight $a^\star$. If specifically $a^\star(\Psi) = \Psi$, Eqs. 1.18 and 1.7 can be connected by

$$\tilde{\vartheta} = c(\mu, \Psi)\cdot\Psi\,, \quad b^\star(\tilde{\vartheta}) = -a(\mu, \Psi)\cdot\Psi\,, \quad \text{and} \quad c^\star(y, \Psi) = b(y, \Psi)\,.$$

Correspondingly, the relationship between Eqs. 1.1 and 1.18 is given by

$$\tilde{\vartheta} = \vartheta\cdot\Psi\,, \quad b^\star(\tilde{\vartheta}) = d(\vartheta, \Psi)\cdot\Psi\,, \quad \text{and} \quad c^\star(y, \Psi) = b(y, \Psi)\,.$$

Mean and variance can be obtained by

$$\mu(\theta) = \mu = \mathbb{E}(y) = \frac{\partial b^\star(\tilde{\vartheta})}{\partial\tilde{\vartheta}}\,, \quad \text{and} \quad \Sigma = \Psi\cdot\frac{\partial^2 b^\star(\tilde{\vartheta})}{\partial\tilde{\vartheta}^2} = \Psi\cdot\frac{\partial\mu}{\partial\tilde{\vartheta}} = \Psi\cdot h(\mu)\,.$$

In this parameterization, the variance Σ is a product of the nuisance parameter Ψ and the function $h(\mu)$. Ψ is termed the dispersion parameter, and h is termed the variance function. In the more general case, Ψ is replaced by

$a^{\star}(\Psi)$ and called the weight function. A detailed discussion of GLM can be found in Chapt. 3.

Chapter 2
The quadratic exponential family

Various generalized estimating equations of order 2 (GEE2) to simultaneously estimate the mean and the association structure can be obtained from the pseudo maximum likelihood 2 (PML2) method. PML2 estimation has been introduced by Gourieroux et al. (1984b), and it is based on the quadratic exponential family. In the first section of this chapter, the quadratic exponential family is introduced with a vectorized form of the association parameters. This representation allows the properties of the linear exponential family to transfer to the quadratic exponential family. It also permits a simple derivation and formulation of the GEE2 (Chapt. 7). Selected properties of the quadratic exponential family are derived in Sect. 2.2. Examples illustrate the applicability of the quadratic exponential family (Sect. 2.3). In the final section of this chapter, we discuss the formulation of the joint distribution of dichotomous variables because of its complexity and its importance for applications.

2.1 Definition

Several GEE2 can be derived from PML2 estimation, and the latter is based on the quadratic exponential family. The important difference between the definitions of the linear exponential family (Chapt. 1) and the quadratic exponential family is that some nuisance parameters $\boldsymbol{\Psi}$ are used in the definition of the linear exponential family, while the covariance matrix $\boldsymbol{\Sigma}$ is used for the quadratic exponential family.

Definition 2.1 (Quadratic exponential family). Let $\boldsymbol{y} \in \mathrm{I\!R}^T$ be a random vector and $\boldsymbol{w} = (y_1^2, y_1 y_2, \ldots, y_1 y_T, y_2^2, y_2 y_3, \ldots, y_T^2)'$, let $\boldsymbol{\mu} \in \Delta \subset \mathrm{I\!R}^T$ be the corresponding mean vector, and let $\boldsymbol{\Sigma}$ be the respective positive definite $T \times T$ covariance matrix. Furthermore, let $a : \mathrm{I\!R}^T \times \mathrm{I\!R}^{T\times T} \to \mathrm{I\!R}$, $b : \mathrm{I\!R}^T \to \mathrm{I\!R}$, $\boldsymbol{c} : \mathrm{I\!R}^T \times \mathrm{I\!R}^{T\times T} \to \mathrm{I\!R}^T$, and $\boldsymbol{j} : \mathrm{I\!R}^T \times \mathrm{I\!R}^{T\times T} \to \mathrm{I\!R}^{T(T+1)/2}$

be (measurable) functions. The T-dimensional quadratic exponential family with mean $\boldsymbol{\mu}$ and covariance matrix $\boldsymbol{\Sigma}$ is given by the set of distributions with density functions

$$f(\boldsymbol{y}||\boldsymbol{\mu},\boldsymbol{\Sigma}) = \exp\Big(\boldsymbol{c}(\boldsymbol{\mu},\boldsymbol{\Sigma})'\boldsymbol{y} + a(\boldsymbol{\mu},\boldsymbol{\Sigma}) + b(\boldsymbol{y}) + \boldsymbol{j}(\boldsymbol{\mu},\boldsymbol{\Sigma})'\boldsymbol{w}\Big)\,. \tag{2.1}$$

By letting $\boldsymbol{\vartheta} = \boldsymbol{c}(\boldsymbol{\mu},\boldsymbol{\Sigma})$ and $\boldsymbol{\lambda} = \boldsymbol{j}(\boldsymbol{\mu},\boldsymbol{\Sigma})$ as in Sect. 1.3, this density may be rewritten as

$$f(\boldsymbol{y}||\boldsymbol{\vartheta},\boldsymbol{\lambda}) = \exp\Big(\boldsymbol{\vartheta}'\boldsymbol{y} - d(\boldsymbol{\vartheta},\boldsymbol{\lambda}) + b(\boldsymbol{y}) + \boldsymbol{\lambda}'\boldsymbol{w}\Big)\,. \tag{2.2}$$

Remark 2.2.

- The representation of the quadratic exponential family in Eq. 2.2 does not immediately open up the term quadratic. However, the function $\boldsymbol{j}(\boldsymbol{\mu},\boldsymbol{\Sigma})'\boldsymbol{w}$ can also be represented by the quadratic form $\boldsymbol{y}'\boldsymbol{D}\boldsymbol{y}$ for a symmetric matrix $\boldsymbol{D}(\boldsymbol{\mu},\boldsymbol{\Sigma})$ because

$$\boldsymbol{y}'\boldsymbol{D}\boldsymbol{y} = \sum_{t=1}^{T}\sum_{t'=1}^{T} y_t y_{t'}[\boldsymbol{D}]_{tt'} = \sum_{t=1}^{T} y_t^2[\boldsymbol{D}]_{tt} + 2\sum_{t'>t} y_t y_{t'}[\boldsymbol{D}]_{tt'} = \boldsymbol{j}'\boldsymbol{w}\,,$$

  and $\boldsymbol{j} = ([\boldsymbol{D}]_{11}, 2[\boldsymbol{D}]_{12}, \ldots, 2[\boldsymbol{D}]_{1T}, [\boldsymbol{D}]_{22}, 2[\boldsymbol{D}]_{23}, \ldots, [\boldsymbol{D}]_{TT})'$. Therefore, the exponential family is quadratic because the exponent can be formulated quadratic in $\boldsymbol{y}$ with coefficient matrix $\boldsymbol{D}(\boldsymbol{\mu},\boldsymbol{\Sigma})$. This quadratic form of the quadratic exponential family has been used by Gourieroux et al. (1984b). However, the vectorized version is more convenient for deriving some properties of the quadratic exponential family (see next section).
- In contrast to the linear exponential family (Definition 1.1), where second-order moments were treated as nuisance, these are of interest in the quadratic exponential family. Moments of order three and higher are, however, ignored and set to $\boldsymbol{0}$. This is important for the definition of the joint distribution of a T-variate dichotomous random vector (Sect. 2.4). If readers are interested in higher order exponential families, they may refer, e.g., to Holly et al. (2008).
- We can define the vector $\boldsymbol{\nu} = (\nu_{11}, \nu_{12}, \ldots, \nu_{22}, \ldots)'$ in analogy to $\boldsymbol{\lambda}$ with $\nu_{tt'} = \sigma_{tt'} + \mu_t\mu_{t'}$, where $\nu_{tt'} = \mathbb{E}(y_t y_{t'})$ denotes the second ordinary moment, and $\sigma_{tt'} = [\boldsymbol{\Sigma}]_{tt'}$ is the tt'th element of the covariance matrix $\boldsymbol{\Sigma}$. $\boldsymbol{\nu}$ will be used below to formulate GEE using the second ordinary moments, which are also termed product moments.

2.2 Selected properties

By condensing $\boldsymbol{\vartheta}$ and $\boldsymbol{\lambda}$ to one column vector and $\boldsymbol{y}$ and $\boldsymbol{w}$ to another column vector, the properties of Theorem 1.2 for the linear exponential family can be extended to the quadratic exponential family. For example, mean and variance of the quadratic exponential family with vectorized association parameter $\boldsymbol{\lambda}$ are given by

$$\mathbb{E}\big((\boldsymbol{y}', \boldsymbol{w}')'\big) = \frac{\partial d(\boldsymbol{\vartheta}, \boldsymbol{\lambda})}{\partial(\boldsymbol{\vartheta}', \boldsymbol{\lambda}')'} = \begin{pmatrix} \boldsymbol{\mu} \\ \boldsymbol{\nu} \end{pmatrix}, \qquad \text{and}$$

$$\mathbb{V}\mathrm{ar}\Big((\boldsymbol{y}', \boldsymbol{w}')'\Big) = \begin{pmatrix} \mathbb{V}\mathrm{ar}(\boldsymbol{y}) & \mathbb{C}\mathrm{ov}(\boldsymbol{y}, \boldsymbol{w}) \\ \mathbb{C}\mathrm{ov}(\boldsymbol{w}, \boldsymbol{y}) & \mathbb{V}\mathrm{ar}(\boldsymbol{w}) \end{pmatrix} = \frac{\partial(\boldsymbol{\mu}', \boldsymbol{\nu}')}{\partial(\boldsymbol{\vartheta}', \boldsymbol{\lambda}')'} = \begin{pmatrix} \frac{\partial \boldsymbol{\mu}}{\partial \boldsymbol{\vartheta}'} & \frac{\partial \boldsymbol{\nu}}{\partial \boldsymbol{\vartheta}'} \\ \frac{\partial \boldsymbol{\mu}}{\partial \boldsymbol{\lambda}'} & \frac{\partial \boldsymbol{\nu}}{\partial \boldsymbol{\lambda}'} \end{pmatrix}. \tag{2.3}$$

Similarly, Property 1.6, which has been given for the linear exponential family, can be generalized to the quadratic exponential family.

Property 2.3. For any $\boldsymbol{\mu}, \boldsymbol{\mu}_0 \in \Delta$ and for all positive definite $T \times T$ covariance matrices $\boldsymbol{\Sigma}$ and $\boldsymbol{\Sigma}_0$,

$$\begin{aligned} &\boldsymbol{c}(\boldsymbol{\mu}, \boldsymbol{\Sigma})'\boldsymbol{\mu}_0 + a(\boldsymbol{\mu}, \boldsymbol{\Sigma}) + \boldsymbol{j}(\boldsymbol{\mu}, \boldsymbol{\Sigma})'\boldsymbol{\nu}_0 \\ &\qquad \le \boldsymbol{c}(\boldsymbol{\mu}_0, \boldsymbol{\Sigma}_0)'\boldsymbol{\mu}_0 + a(\boldsymbol{\mu}_0, \boldsymbol{\Sigma}_0) + \boldsymbol{j}(\boldsymbol{\mu}_0, \boldsymbol{\Sigma}_0)'\boldsymbol{\nu}_0 . \end{aligned} \tag{2.4}$$

Equality holds, if and only if $\boldsymbol{\mu} = \boldsymbol{\mu}_0$ and $\boldsymbol{\Sigma} = \boldsymbol{\Sigma}_0$.

Proof. Analogously to Property 1.6, Property 2.3 is a direct consequence of Kullback's inequality. □

2.3 Examples for quadratic exponential families

In this section, we give three examples for quadratic exponential families. We start with the univariate normal distribution because the normal distribution is of great importance for PML2 estimation. It is followed by the multivariate normal distribution, and the last example of this section is one given by Gourieroux et al. (1984b). In the next section, two further examples are given. The reader should note that some standard distributions that belong to two-parameter linear exponential families (see, e.g., Example 1.11) do not belong to the class of quadratic exponential families. A typical example is the Gamma distribution.

Example 2.4 (Univariate normal distribution). The density of the univariate normal distribution has been given in Example 1.10. If Ψ is replaced by $\Sigma = \sigma^2$, one obtains $b(y) = 0$ as well as

$$c(\mu, \Sigma) = \frac{\mu}{\Sigma}\,, a(\mu, \Sigma) = -\frac{1}{2}\frac{\mu^2}{\Sigma} - \tfrac{1}{2}\ln(2\pi\Sigma)\,, \quad \text{and} \quad j(\mu, \Sigma) = -\frac{1}{2}\frac{1}{\Sigma}\,.$$

Example 2.5 (Multivariate normal distribution). The classical example for a distribution belonging to the quadratic exponential family is the T-dimensional normal distribution. Its density has been given in Example 1.12. By replacing $\boldsymbol{\Psi}$ with $\boldsymbol{\Sigma}$, we obtain $b(\boldsymbol{y}) = 0$,

$$\boldsymbol{j}(\boldsymbol{\mu}, \boldsymbol{\Sigma}) = -\tfrac{1}{2}\left([\boldsymbol{\Sigma}^{-1}]_{11}, 2[\boldsymbol{\Sigma}^{-1}]_{12}, \ldots, 2[\boldsymbol{\Sigma}^{-1}]_{1T}, [\boldsymbol{\Sigma}^{-1}]_{22}, \ldots, [\boldsymbol{\Sigma}^{-1}]_{TT}\right),$$
$$a(\boldsymbol{\mu}, \boldsymbol{\Sigma}) = -\tfrac{1}{2}\boldsymbol{\mu}'\boldsymbol{\Sigma}^{-1}\boldsymbol{\mu} - \tfrac{T}{2}\ln(2\pi) - \tfrac{1}{2}\ln\left(\det(\boldsymbol{\Sigma})\right), \quad \text{and} \quad \boldsymbol{c}(\boldsymbol{\mu}, \boldsymbol{\Sigma}) = \boldsymbol{\Sigma}^{-1}\boldsymbol{\mu}\,.$$

Example 2.6 (Discrete distribution on {-1, 0, 1}). Consider a discrete distribution on $\{-1, 0, 1\}$ with probabilities p_{-1}, p_0, and p_1. Its density is given by

$$\overset{\approx}{f}(y||p_{-1}, p_0, p_1) = p_{-1}^{y(y-1)/2}\, p_0^{(1-y^2)}\, p_1^{y(1+y)/2}\,,$$

subject to $p_{-1} + p_0 + p_1 = 1$, $p_1 - p_{-1} = \mu$, and $p_{-1} + p_1 = \Sigma + \mu^2$. We obtain

$$f(y||\mu, \Sigma) = \exp\left(\frac{y}{2}\ln\left(\frac{\Sigma + \mu^2 + \mu}{\Sigma + \mu^2 - \mu}\right) + \frac{y^2}{2}\ln\left(\frac{(\Sigma + \mu^2 + \mu)(\Sigma + \mu^2 - \mu)}{4 \cdot (1 - \Sigma - \mu^2)^2}\right)\right.$$
$$\left. + \ln\left(1 - \Sigma - \mu^2\right)\right).$$

We complete the example by noting that $a(\mu, \Sigma) = \ln\left(1 - \Sigma - \mu^2\right)$, $b(y) = 0$, $c(\mu, \Sigma) = \frac{1}{2}\ln\dfrac{\Sigma + \mu^2 + \mu}{\Sigma + \mu^2 - \mu}$, and $j(\mu, \Sigma) = \frac{1}{2}\ln\dfrac{(\Sigma + \mu^2 + \mu)(\Sigma + \mu^2 - \mu)}{4 \cdot (1 - \Sigma - \mu^2)^2}$.

2.4 The joint distribution of dichotomous random variables

Several representations of the joint distribution of T dichotomous random variables have been given in the literature, and their pros and cons have been extensively discussed (Kauermann, 1997; Liang et al., 1992; Prentice, 1988). In this section, we first consider the joint distribution of two dichotomous random variables and show that this distribution is a member of the quadratic exponential family. Furthermore, we consider the joint distribution of T dichotomous random variables and embed a specific version of it into the quadratic exponential family. In this version, all moments of order three and above are set to 0.

2.4.1 *The joint distribution of two dichotomous random variables*

The joint distribution of two dichotomous items is often used in applications. The standard parameter to measure the association between the responses is the odds ratio (OR). We therefore define the OR first and next give the definition of the joint distribution. Desirable properties and interpretations of the OR and other measures of association for pairs of binary responses have been described in detail, e.g., by Bishop et al. (1975).

The starting point is the 2×2 Table 2.1, which displays a bivariate random vector $\boldsymbol{y} = (y_1, y_2)'$, where $\pi_{tt'} = \mathbb{P}(y_1 = t, y_2 = t')$. The means are given by $\mu_1 = \pi_1 = \mathbb{P}(y_1 = 1)$ and $\mu_2 = \pi_2 = \mathbb{P}(y_2 = 1)$. For simplicity, we assume that all probabilities are > 0 and < 1.

Table 2.1 2×2 table for y_1 and y_2

		y_2 = 0	y_2 = 1	
y_1	0	π_{00}	π_{01}	$1 - \pi_1$
	1	π_{10}	π_{11}	π_1
		$1 - \pi_2$	π_2	1

Definition 2.7 (Odds, odds ratio).
1. The odds of column 1 is $O_1 = \frac{\pi_{00}}{\pi_{10}}$, and the odds of column 2 is $O_2 = \frac{\pi_{01}}{\pi_{11}}$.
2. The odds ratio (OR) τ, also termed cross product ratio, is the ratio of the two odds
$$\tau_{12} = \mathrm{OR}_{12} = \mathrm{OR}(y_1, y_2) = \frac{O_1}{O_2} = \frac{\pi_{00}\,\pi_{11}}{\pi_{01}\,\pi_{10}}\,.$$

Definition 2.8 (Joint distribution of two dichotomous random variables). The joint distribution of two dichotomous items y_1 and y_2 is given by

$$\begin{aligned}\mathbb{P}(\boldsymbol{y}) = \mathbb{P}(y_1, y_2) = \exp\big(&y_1 y_2 \ln \pi_{11} + y_1(1 - y_2) \ln \pi_{10} \\ &+ (1 - y_1) y_2 \ln \pi_{01} + (1 - y_1)(1 - y_2) \ln \pi_{00}\big)\,.\end{aligned} \tag{2.5}$$

Remark 2.9. Equation 2.5 can be rewritten as

$$\begin{aligned}\mathbb{P}(\boldsymbol{y}) = \exp\big(&y_1 y_2 [\ln(\pi_{11}\pi_{00}) - \ln(\pi_{10}\pi_{01})] \\ &+ y_1[\ln \pi_{10} - \ln \pi_{00}] + y_2[\ln \pi_{01} - \ln \pi_{00}] + \ln \pi_{00}\big)\,,\end{aligned}$$

and it can therefore be seen that the joint distribution of two dichotomous random variables belongs to the quadratic exponential family. The natural parameters have a loglinear representation and are given by

$$\begin{aligned}\vartheta_1 &= \operatorname{logit}\{\mathbb{P}(y_1 = 1|y_2 = 0)\} = \ln \pi_{10} - \ln \pi_{00}\,, \\ \vartheta_2 &= \operatorname{logit}\{\mathbb{P}(y_2 = 1|y_1 = 0)\} = \ln \pi_{01} - \ln \pi_{00}\,, \quad \text{and} \\ \lambda_{12} &= \log \mathrm{OR}(y_1, y_2) = \ln(\pi_{11}\pi_{00}) - \ln(\pi_{10}\pi_{01})\,.\end{aligned} \tag{2.6}$$

The natural parameters ϑ_t are conditional probabilities, and λ_{12} is an unconditional log OR. The normalization constant is $d(\boldsymbol{\vartheta}, \boldsymbol{\lambda}) = -\ln \pi_{00}$, and $b(\boldsymbol{y}) = 0$. The parameters λ_{11} and λ_{22} of y_1^2 and y_2^2 are identical to 0 because the variances are completely specified by the means.

Finally, in Eq. 2.6 the joint distribution is formulated using conditional logits. In the next section, a representation in the marginal moments $(\pi_1, \pi_2, \lambda_{12})'$ will be considered.

To derive GEE for dichotomous items using the quadratic exponential family, we require the following functional relationship between the second ordinary moments $\pi_{11} = \mathbb{E}(y_1 y_2)$ and the OR τ_{12} in Chapt. 7. This functional relationship has been given, e.g., by Bishop et al. (1975):

$$\mathbb{E}(y_1 y_2) = \pi_{11} = \begin{cases} \dfrac{f_{12} - \sqrt{f_{12}^2 - 4\tau_{12}(\tau_{12} - 1)\pi_1\pi_2}}{2(\tau_{12} - 1)} & \text{if} \quad \tau_{12} \neq 1\,, \\ \tau_{12}\pi_1\pi_2 & \text{if} \quad \tau_{12} = 1\,, \end{cases} \tag{2.7}$$

where $f_{12} = \big(1 - (1 - \tau_{12})(\pi_1 + \pi_2)\big)$. By using

$$\tau_{12} = \frac{\pi_{11}(1 - \pi_1 - \pi_2 + \pi_{11})}{(\pi_1 - \pi_{11})(\pi_2 - \pi_{11})} \quad \text{and} \quad \varrho = \frac{\pi_{11} - \pi_1\pi_2}{\sqrt{\pi_1(1 - \pi_1)\pi_2(1 - \pi_2)}}\,,$$

a one-to-one functional relation between the OR τ and the correlation coefficient $\varrho = \mathbb{C}\mathrm{orr}(y_1, y_2)$ can also be established. As a result, the OR can therefore also be written as a function of the means and the correlation coefficient, and the correlation coefficient can be written as a function of the means and the OR.

2.4.2 The joint distribution of T dichotomous random variables

The joint distribution of two binary random variables can be extended easily to T dichotomous random variables. Under the restrictive assumptions that all third and higher order moments equal 0, this distribution also belongs to the quadratic exponential family. Three different parameterizations are often used in practice for this model, the loglinear parameterization, the marginal parameterization using contrasts of odds ratios, and the marginal parameterization using the correlation coefficient as the measure of association.

Definition 2.10. Consider T dichotomous random variables $y_1, \ldots, y_T$, which are summarized to a vector $\boldsymbol{y}$. The joint distribution in the loglinear parameterization is given by

$$\mathbb{P}(\boldsymbol{y}) = \exp\Big(\sum_{t=1}^{T} y_t\vartheta_t + \sum_{t<t'} y_t y_{t'}\lambda_{tt'} + \sum_{t<t'<t''} y_t y_{t'} y_{t''}\zeta_{tt't''} + \ldots + y_1 y_2 \cdots y_T \zeta_{1\ldots T} - d(\boldsymbol{\vartheta}, \boldsymbol{\lambda}, \boldsymbol{\zeta})\Big), \quad (2.8)$$

for $\boldsymbol{\zeta}' = (\zeta_{tt't''}, \ldots, \zeta_{1\ldots T})$ with ϑ_t being logits of conditional probabilities

$$\vartheta_t = \operatorname{logit}\{\mathbb{P}(y_t = 1 | y_{t'} = 0, t \neq t'\} = \ln \frac{\mathbb{P}(y_t = 1 | y_{t'} = 0, t \neq t')}{\mathbb{P}(y_t = 0 | y_{t'} = 0, t \neq t')} \,.$$

The second-order moments are conditional log ORs

$$\begin{aligned}\lambda_{tt'} &= \log \mathrm{OR}(y_t, y_{t'} | y_{t''} = 0) \\ &= \ln \frac{\mathbb{P}(y_t = 1, y_{t'} = 1 | y_{t''} = 0)}{\mathbb{P}(y_t = 0, y_{t'} = 1 | y_{t''} = 0)} - \ln \frac{\mathbb{P}(y_t = 1, y_{t'} = 0 | y_{t''} = 0)}{\mathbb{P}(y_t = 0, y_{t'} = 0 | y_{t''} = 0)}\end{aligned}$$

with $t'' \neq t, t'$.

Remark 2.11.

- In Definition 2.10, second-order moments are conditional log ORs, while they were unconditional in Eq. 2.6.
- If all parameters of order three and above are set to 0, thus $\boldsymbol{\zeta} = \boldsymbol{0}$, the joint distribution of T binary random variables belongs to the quadratic exponential family.

An alternative parameterization of the joint distribution of T dichotomous items is in terms of marginal rather than fully conditional distributions. In applications, the first two moments are of primary interest, and these are initially specified by

$$\mathbb{P}(y_t = 1) = \mu_t, \qquad \text{and} \qquad \mathrm{OR}(y_t, y_{t'}) = \tau_{tt'} \,. \quad (2.9)$$

The parameterization of the full joint distribution can be completed in several ways, e.g., in terms of contrasts of conditional ORs (Liang et al., 1992):

$$\begin{aligned}
\zeta_{tt't''} &= \ln\mathrm{OR}(y_t, y_{t'}|y_{t''}=1) - \ln\mathrm{OR}(y_t, y_{t'}|y_{t''}=0)\,, \qquad (2.10)\\
\zeta_{rr'tt'} &= \ln\mathrm{OR}(y_t, y_{t'}|y_r=1, y_{r'}=1) - \ln\mathrm{OR}(y_t, y_{t'}|y_r=1, y_{r'}=0)\\
&\quad - \ln\mathrm{OR}(y_t, y_{t'}|y_r=0, y_{r'}=1) + \ln\mathrm{OR}(y_t, y_{t'}|y_r=0, y_{r'}=0)\,,\\
\zeta_{t_1\ldots t_S} &= \sum(-1)^{\sum_{s=3}^{S} y_{t_s}+S-2}\ln\mathrm{OR}(y_{t_1}, y_{t_2}|y_{t_3},\ldots,y_{t_S})\,.
\end{aligned}$$

Here, the sum is taken over the 2^{S-2} possible combinations of $(y_{t_3},\ldots,y_{t_S})$. Another completion of the joint distribution uses the following higher order moments:

$$\varsigma_{tt't''} = \ln\frac{\pi_{111}\ \pi_{000}\ \pi_{100}\ \pi_{010}}{\pi_{101}\ \pi_{011}\ \pi_{110}\ \pi_{000}}\,, \qquad (2.11)$$

$$\varsigma_{rr'tt'} = \ln\frac{\pi_{1111}\ \pi_{0011}\ \pi_{1100}\ \pi_{0000}\ \pi_{1010}\ \pi_{0110}\ \pi_{1001}\ \pi_{0101}}{\pi_{1110}\ \pi_{1101}\ \pi_{1011}\ \pi_{0111}\ \pi_{1000}\ \pi_{0100}\ \pi_{0010}\ \pi_{0001}}\,, \qquad (2.12)$$

$$\begin{aligned}
\varsigma_{t_1\ldots t_S} &= (-1)^S\ln\prod\nolimits_{\left\{(t_1,\ldots,t_S):\sum_{s=1}^{S} t_s=2n,\ n\in\mathbb{N}\right\}}\pi_{t_1\ldots t_S} + \qquad (2.13)\\
&\quad (-1)^{S+1}\ln\prod\nolimits_{\left\{(t_1,\ldots,t_S):\sum_{s=1}^{S} t_s=2n+1,\ n\in\mathbb{N}\right\}}\pi_{t_1\ldots t_S}\,.
\end{aligned}$$

Kauermann (1997) has shown that the loglinear and the marginal parameterization have a 1:1 correspondence, although higher order marginal and loglinear parameters differ. The most important difference between the loglinear and marginal parameters is in the interpretation of the first two moments. In the loglinear parameterization, they can be interpreted as conditional probabilities of y_t given $y_{t'}$, or as conditional log ORs of $y_t, y_{t'}$ given $y_{t''}$.

The pros and cons of both parameterizations have been discussed by Liang et al. (1992). The major advantage of the marginal model over the loglinear model is its reproducibility. Thus, if $\boldsymbol{y}$ satisfies a marginal model, then any subset of $\boldsymbol{y}$ also does. Hence, the interpretation of the marginal parameters is independent of the length T of $\boldsymbol{y}$. One drawback of the marginal parameterization is that extensive computations are required for parameter estimation, see, e.g., Fitzmaurice and Laird (1993). Even more important, the parameter space is restricted (see next section), and this restriction has immediate consequences for choosing the weight matrices for GEEs with dichotomous dependent variables (see Sect. 7.3.4).

Another formulation of the joint distribution is based on correlations. Specifically, Bahadur (1961) has shown that the joint distribution of T dichotomous random variables can be written as

$$\begin{aligned}
\mathbb{P}(y_1,\ldots,y_T) &= \prod_{t=1}^{T}\Big(\pi_t^{y_t}(1-\pi_t)^{1-y_t}\Big)\Big(1+\sum_{t<t'}\varrho_{tt'}z_t z_{t'} \qquad (2.14)\\
&\quad + \sum_{t<t'<t''}\varrho_{tt't''}z_t z_{t'} z_{t''} + \cdots + \varrho_{1\ldots T}z_1 z_2\ldots z_T\Big)\,,
\end{aligned}$$

where $z_i = (y_i - \mu_i)/\sigma_i$ is the standardized variable with $\sigma_i = \sqrt{\mu_i(1-\mu_i)}$, $\varrho_{tt'} = \mathbb{C}orr(y_t, y_{t'}) = \mathbb{E}(z_t z_{t'})$, $\varrho_{tt't''} = \mathbb{E}(z_t z_{t'} z_{t''})$, etc. The Bahadur representation is similar to the representation in the marginal OR, and the corresponding parameter space is restricted, too.

2.4.3 Restriction of the parameter space in marginal models

When using the parameterization in the correlation coefficient, the parameter space is restricted for $T \geq 2$. Consider the 2×2 Table 2.1 with means $\pi_1 = \pi_{10} + \pi_{11}$ and $\pi_2 = \pi_{01} + \pi_{11}$. Then, the probability π_{11} is restricted to

$$\max(0, \pi_1 + \pi_2 - 1) \leq \pi_{11} \leq \min(\pi_1, \pi_2)\,.$$

Because the correlation coefficient

$$\varrho = \frac{\pi_{11} - \pi_1\pi_2}{\sqrt{\pi_1(1-\pi_1)\cdot\pi_2(1-\pi_2)}}$$

directly depends on the restricted π_{11}, ϱ is also restricted. Specifically, ϱ is constrained by (Prentice, 1988)

$$\max\left\{-\sqrt{\frac{\pi_1\pi_2}{\varpi_1\varpi_2}}, -\sqrt{\frac{\varpi_1\varpi_2}{\pi_1\pi_2}}\right\} \leq \varrho \leq \min\left\{\sqrt{\frac{\pi_1\varpi_2}{\varpi_1\pi_2}}, \sqrt{\frac{\varpi_1\pi_2}{\pi_1\varpi_2}}\right\}, \qquad (2.15)$$

where $\varpi_t = 1 - \pi_t$. Similar restrictions hold for moments $\varrho_{t_1 \ldots t_s}$ of order three and above.

Examples for the restrictions are given in Table 2.2. Extreme restrictions occur when one probability is approaching the boundary of the parameter space, while the other is approximately 0.5. For example, if $\pi_1 = 0.01$ and $\pi_2 = 0.5$, ϱ is bounded by ± 0.1. It is bounded by ± 0.03 if π_1 is 0.001.

Table 2.2 Restrictions of the correlation coefficient in a 2×2 setting given the marginal means π_1 and π_2

		π_2		
		0.1	0.3	0.5
	0.1	-0.11;1.00	-0.22;0.51	-0.33;0.33
π_1	0.3		-0.43;1.00	-0.65;0.65
	0.5			-1.00;1.00

Marginal moments other than the correlation coefficient are also restricted. Specifically, the OR is restricted for binary dependent variables if $T \geq 3$. For example, let the parameters $(\pi_1, \pi_2, \pi_3, \tau_{12}, \tau_{13})$ be fixed, and $T = 3$. $\mathbb{E}(y_1 y_2)$

and $\mathbb{E}(y_1 y_3)$ can be determined through these parameters. For example, the domain of $\mathbb{E}(y_2 y_3)$ is restricted to

$$\max\left\{0, \mathbb{E}(y_1 y_2) + \mathbb{E}(y_1 y_3) - \pi_1\right\} \le \mathbb{E}(y_2 y_3) \le \min\left\{\pi_2, \pi_3\right\}. \tag{2.16}$$

As a result, the domain of τ_{23} is also restricted because

$$\tau_{23} = \frac{\mathbb{E}(y_2 y_3)\big(1 - \pi_2 - \pi_3 + \mathbb{E}(y_2 y_3)\big)}{\big(\pi_2 - \mathbb{E}(y_2 y_3)\big)\big(\pi_3 - \mathbb{E}(y_2 y_3)\big)}.$$

Analogous restrictions exist for moments of order greater than three.

In summary, the OR parameterization can be used without any constraints for two dichotomous random variables, while the parameter space of the correlation coefficient is already restricted in this case. If more than two dichotomous random variables are considered, the parameter space is restricted in all marginal parameterizations. For greater flexibility in the parameter space, higher order moments should be added to analyze correlated dichotomous random variables.

As a last example for the restriction of the parameter space, consider the joint distribution of T binary responses $\boldsymbol{y} = (y_1, \ldots, y_T)'$ in the Bahadur representation (Eq. 2.14) with all moments $\varrho_{t_1 \ldots t_s}$ of three and above being set to 0:

$$\mathbb{P}(\boldsymbol{y}) = \prod_{t=1}^{T} \left(\pi_t^{y_t} (1 - \pi_t)^{1 - y_t}\right)\left(1 + \sum_{t<t'} \varrho_{tt'} \frac{(y_t - \pi_t)}{\sqrt{\pi_t(1 - \pi_t)}} \frac{(y_{t'} - \pi_{t'})}{\sqrt{\pi_{t'}(1 - \pi_{t'})}}\right).$$

For simplicity, let $\pi_t = \pi > 0$, $\varrho_{tt'} = \varrho$, $T = 2k$, $k \in \mathbb{N}$. Then, a sequence of m zeros and m ones has probability (Prentice, 1988)

$$\pi^m (1 - \pi)^m \left(1 + \varrho\left(\tfrac{1}{2} m(m-1)\tfrac{1-\pi}{\pi} + \tfrac{1}{2} m(m-1)\tfrac{\pi}{1-\pi} - m^2\right)\right),$$

yielding $\varrho \le |m^2 - \frac{1}{2}m(m-1)\frac{1-\pi}{\pi} - \frac{1}{2}m(m-1)\frac{\pi}{1-\pi}|^{-1}$. For example, $\varrho \le |\frac{1}{m}|$ if $\pi = \frac{1}{2}$. Thus, the impact of the restrictions increases with the number of observations T per cluster.

A summary of the permissible ranges of dependent dichotomous variables has been given by Chaganty and Deng (2007). The effect of ignoring the bounds has been nicely illustrated by Sabo and Chaganty (2010). A detailed discussion how these restrictions may be overcome can be found in Ziegler and Vens (2010); also see Shults (2011).

Chapter 3
Generalized linear models

In this chapter, the class of generalized linear models (GLM) will be introduced as required for understanding the idea of generalized estimating equations (GEE). Univariate GLMs are considered first, followed by multivariate GLMs. For an in-depth discussion of GLM, the reader may refer to the literature (Fahrmeir and Tutz, 2001; Hardin and Hilbe, 2007; McCullagh and Nelder, 1989).

3.1 Univariate generalized linear models

3.1.1 Definition

Definition 3.1. Let $\boldsymbol{y} = (y_1, \ldots, y_n)'$ be an n dimensional random vector, let $\boldsymbol{X} = (\boldsymbol{x}_1, \ldots, \boldsymbol{x}_n)'$ be an $n \times p$ matrix of fixed and/or stochastic regressors, $\boldsymbol{\beta} = (\beta_1, \ldots, \beta_p)'$ a p dimensional parameter vector, and $\boldsymbol{\varepsilon} = (\epsilon_1, \ldots, \epsilon_n)'$ an n dimensional random vector of errors. We assume that the pairs $(y_i, \boldsymbol{x}_i)$ are independent and that $y_i|\boldsymbol{x}_i$ are identically distributed for all $i = 1, \ldots, n$. The $p \times p$ matrix $\frac{1}{n}\boldsymbol{X}'\boldsymbol{X}$ is assumed to converge (almost surely) to a non-stochastic regular matrix $\boldsymbol{Q}$ as $n \to \infty$.

In GLMs, the vector of observations $\boldsymbol{y}$ is additively decomposed into a systematic component $\boldsymbol{\mu}$ and an error term $\boldsymbol{\varepsilon}$,

$$\boldsymbol{y} = \boldsymbol{\mu} + \boldsymbol{\varepsilon} \,,$$

where $\boldsymbol{\varepsilon}$ and $\boldsymbol{X}$ are assumed to be stochastically independent, i.e., $\mathbb{E}(\boldsymbol{\varepsilon}|\boldsymbol{X}) = \boldsymbol{0}$, and $\boldsymbol{\mu} = (\mu_1, \ldots, \mu_n)'$ is the vector of conditional means $\mathbb{E}(y_i|\boldsymbol{x}_i) = \mu_i$ of y_i given $\boldsymbol{x}_i$.

In a univariate GLM, the conditional density $f(y_i||\vartheta_i) = f_{y_i|\boldsymbol{x}_i}(y_i||\vartheta_i)$ belongs to the univariate linear exponential family with natural parameter

ϑ_i. Furthermore, the conditional mean $\mu_i = \mathbb{E}(y_i|\boldsymbol{x}_i)$ is related to the linear predictor $\eta_i = \boldsymbol{x}_i'\boldsymbol{\beta}$ by a one to one link function $g\colon \mathbb{R} \to \mathbb{R}$, which is assumed to be sufficiently often continuously differentiable: $g(\mu_i) = \eta_i = \boldsymbol{x}_i'\boldsymbol{\beta}$. The inverse g^{-1} of the link function g is termed response function. For simplicity, the following vector and matrix notation is used: $\boldsymbol{\eta} = (\eta_1, \ldots, \eta_n)'$, $\boldsymbol{g}(\boldsymbol{\mu}) = (g(\mu_1), \ldots, g(\mu_n))'$, and $\boldsymbol{g}(\boldsymbol{\mu}) = \boldsymbol{\eta} = \boldsymbol{X}\boldsymbol{\beta}$.

The term generalized linear model is used because the explanatory variables $\boldsymbol{x}_i$ are linearly connected with the parameter of interest $\boldsymbol{\beta}$ to the linear predictor η_i. However, the linear predictor η_i can be connected with the dependent variable y_i in a more general way than through the identity function. Therefore, one often writes $\mu_i = \mu_i(\eta_i) = \mu_i(\boldsymbol{x}_i'\boldsymbol{\beta})$ because the conditional mean μ_i of y_i given $\boldsymbol{x}_i$ depends on η_i. For interpretations of the link function, the reader may refer to the literature (Dobson, 2001; Fahrmeir and Tutz, 2001).

3.1.2 Parameterization and natural link function

In Definition 3.1, a functional relationship between the parameters ϑ_i, $i = 1, \ldots, n$, from the linear exponential family and the parameter of interest $\boldsymbol{\beta}$ has not been established. By use of Eq. 1.6, i.e., $\vartheta_i = c(\mu_i, \boldsymbol{\Psi})$, ϑ_i can be written as a function of $\boldsymbol{\beta}$, i.e., $\vartheta_i = c\big(g^{-1}(\eta_i), \boldsymbol{\Psi}\big) = c\big(\mu_i(\eta_i), \boldsymbol{\Psi}\big)$.

In the special case of $\vartheta_i = \eta_i$, the GLM is called GLM with natural link function. In this case, we have a linear model for $\boldsymbol{\beta}$, i.e., $\boldsymbol{\vartheta} = \boldsymbol{X}\boldsymbol{\beta}$, and the link function $\boldsymbol{g}$ is identical to the function $\boldsymbol{c}$ of the mean structure parameterization from Sect. 1.3.

3.1.3 Examples

Example 3.2 (GLM for continuous data). If $y_i|\boldsymbol{x}_i$ follows a univariate normal distribution with variance σ^2, the classical linear model with stochastic regressors is obtained by choosing the natural link function $g = \text{ident}$, yielding $\mathbb{E}(y_i|\boldsymbol{x}_i) = \mu_i = g^{-1}(\mu_i) = \eta_i = \boldsymbol{x}_i'\boldsymbol{\beta}$.

In various applications, a nonlinear relationship $g(\mu_i) = \eta_i = \boldsymbol{x}_i'\boldsymbol{\beta}$ is more appropriate, e.g., if variance stabilization is of interest. A flexible way to model the response function $g^{-1}(\mu_i) = \eta_i = \boldsymbol{x}_i'\boldsymbol{\beta}$ is the Box–Cox or power transformation (Box and Cox, 1964)

$$\eta_i = \frac{\mu_i^\lambda - 1}{\lambda} = g(\mu_i)\,, \quad \text{yielding} \quad \mu_i = g^{-1}(\eta_i) = \sqrt[\lambda]{\lambda\eta_i + 1}$$

for $\lambda \in \mathbb{Z}\backslash 0$. If $\lambda = 0$, the loglinear function $\eta_i = \ln \mu_i$ is obtained by use of l'Hospitals rule.

More generally, Pregibon (1980) proposed power transformations with a shift parameter $\eta_i = \frac{(\mu_i+\lambda_2)^{\lambda_1}-1}{\lambda_1}$. With $\lambda_1 = 1$ and $\lambda_2 = 1$, one obtains the identity link function.

Alternatively, folded power transformations $\eta_i = \frac{\mu_i^\lambda-(1-\mu_i)^\lambda}{\lambda}$ can be used. Choosing $\lambda = 0$ gives the logit link.

Example 3.3 (Models for dichotomous data). A simple choice for dichotomous dependent variables is the identity link, i.e., $\mathbb{E}(y_i|\boldsymbol{x}_i) = \mu_i = \pi_i = \eta_i = \boldsymbol{x}_i'\boldsymbol{\beta}$. Although this model has a simple interpretation and although parameter estimates can be obtained without relying on iterative algorithms, it has a substantial drawback. The conditional mean μ_i is a probability π_i, and $\boldsymbol{x}_i'\boldsymbol{\beta}$ therefore needs to be bounded to the interval $[0; 1]$ for any vector $\boldsymbol{x}_i$.

This can be achieved by using a strictly monotone distribution function F as response function so that $\mu_i = F(\eta_i) = g^{-1}(\eta_i)$. The most intuitive approach is to choose the distribution function Φ from the standard normal distribution as response function. The resulting model

$$\mathbb{E}(y_i|\boldsymbol{x}_i) = \mathbb{P}(y_i = 1\boldsymbol{x}_i) = \mu_i = \pi_i = \Phi(\boldsymbol{x}_i'\boldsymbol{\beta})$$

is termed the probit model, and the linear predictor $\eta_i = \boldsymbol{x}_i'\boldsymbol{\beta}$ is called probit. The link function is the inverse distribution function of the normal distribution, i.e., $g(\mu_i) = \Phi^{-1}(\mu_i)$.

Although the probit model is the most intuitive and often employed in econometrics, the most common choice in biomedical applications is the logit or the logistic regression model. It is obtained by choosing the logit function as link function, which is the natural link function. Specifically,

$$\vartheta_i = \operatorname{logit}(\mu_i) = \operatorname{logit}\bigl(\mathbb{P}(y_i = 1|\boldsymbol{x}_i)\bigr) = \ln \frac{\mathbb{P}(y_i = 1|\boldsymbol{x}_i)}{\mathbb{P}(y_i = 0|\boldsymbol{x}_i)} = \eta_i = \boldsymbol{x}_i'\boldsymbol{\beta}. \quad (3.1)$$

Equation 3.1 shows that the logistic model is a linear model for the log odds of the response $y_i = 1$. The linear predictor η_i of this model is therefore termed logit. The response function is the expit function, having the distribution function of the logistic distribution:

$$F(x) = \operatorname{expit}(x) = \exp(x)/\bigl(1 + \exp(x)\bigr) = \bigl(1 + \exp(-x)\bigr)^{-1}.$$

Further examples can be found, e.g., in Fahrmeir and Tutz (2001), Hardin and Hilbe (2007), and McCullagh and Nelder (1989). GLMs for binary dependent data can also be derived using threshold models. This will be considered in Sect. 3.1.4.

Example 3.4 (Models for count data). Count data are an important class of dependent variables. In this situation, the mean is restricted to positive real numbers. Therefore, a linear model for the linear predictor $\mu_i = \boldsymbol{x}_i'\boldsymbol{\beta}$ leads to restrictions on $\boldsymbol{\beta}$. Like for dichotomous data, these can be avoided by choosing a nonlinear link function.

If y_i given $\boldsymbol{x}_i$ is assumed to be Poisson distributed with mean μ_i, the log-link $\eta_i = g(\mu_i) = \ln(\mu_i)$ is the natural link function, and the exponential function is the response function $\mu_i = \exp(\eta_i)$. The corresponding models are termed loglinear models.

A second common choice is the square root linear model with $\eta_i = g(\mu_i) = 2\sqrt{\mu_i}$ and inverse $\mu_i = \left(\eta_i/2\right)^2 = \left(\boldsymbol{x}_i'\boldsymbol{\beta}/2\right)^2$. Note that the quadratic response function is not injective. The square root link, however, stabilizes the variance of the square root $\sqrt{y}$ of a Poisson distributed random variable y. This can be seen by a first-order Taylor series around μ, which yields $2\sqrt{y} \approx 2\sqrt{\mu} + \frac{1}{\sqrt{\mu}}(y - \mu)$. The mean of $\sqrt{y}$ is approximately $\sqrt{\mu}$, and its variance is approximately 1/4.

3.1.4 Threshold model for dichotomous dependent data

The GLM for dichotomous dependent data has been introduced in Example 3.3. A distribution function was used as the response function to overcome the range restriction of the conditional mean of the dependent variable. The GLM for dichotomous dependent variables can also be derived as threshold models, and they are valuable for interpreting results of regression models for longitudinal binary dependent data.

Consider a linear model for a latent continuous variable $y_i^\star$:

$$y_i^\star = \boldsymbol{x}_i'\boldsymbol{\beta} + \sigma\delta_i = \beta_1 + \boldsymbol{x}_i^{\star\prime}\boldsymbol{\beta}^\star ,$$

where σ is a scale parameter, and δ_i is distributed according to $F(\cdot)$. A common choice is the distribution function of the standard normal distribution, i.e., $F = \Phi$. Furthermore, the model includes a regression constant, i.e., $x_{i1} = 1$ for all i.

The dichotomous variable y_i is observable, and it is connected to the latent variable $y_i^\star$ by a simple threshold relation:

$$y_i = \begin{cases} 0, & \text{if} \quad y_i^\star \leq \tau \\ 1, & \text{if} \quad y_i^\star > \tau \end{cases} .$$

One therefore obtains

$$\mathbb{P}(y_i = 1|\boldsymbol{x}_i) = F\left(\frac{\tau - \beta_0 - \boldsymbol{x}_i^{\star\prime}\boldsymbol{\beta}^\star}{\sigma}\right) .$$

The parameters $\tau, \beta_0, \boldsymbol{\beta}_\star$, and σ are identifiable only when two restrictions are introduced. Usually, $\tau = 0$ and $\sigma^2 = 1$ are chosen. This means that the regression parameters are identified up to a scale parameter σ. Furthermore, the regression constant cannot be identified if the threshold parameter is unknown. With these restrictions, we obtain

$$\mathbb{P}(y_i = 1|\boldsymbol{x}_i) = F(\beta_0 + \boldsymbol{x}_i^{\star'}\boldsymbol{\beta}^\star) = \mu_i\,.$$

Thus, $\mu_i = F(\beta_0 + \boldsymbol{x}_i^{\star'}\boldsymbol{\beta}^\star)$ is the mean of y_i given $\boldsymbol{x}_i$ if the error of the linear model for the latent variable is distributed according to $F(\cdot)$, and the model on the observable level therefore is $y_i = \mu_i + \epsilon_i$.

3.2 Multivariate generalized linear models

3.2.1 Definition

Definition 3.5. Consider n stochastic vectors $\boldsymbol{y}_1, \ldots, \boldsymbol{y}_n$ of length $T \times 1$. $\boldsymbol{X}_1, \ldots, \boldsymbol{X}_n$ are the corresponding $T \times p$ fixed or stochastic matrices of regressors. Let $(\boldsymbol{y}_i, \boldsymbol{X}_i)$ be independently identically distributed (i.i.d.), and $\mathbb{E}(\varepsilon_i|\boldsymbol{X}_j) = \boldsymbol{0}$ for all i, j. Finally, assume that the matrix $\frac{1}{n}\sum\limits_{i=1}^{n}\boldsymbol{X}_i'\boldsymbol{X}_i$ converges to a non-stochastic regular matrix $\boldsymbol{Q}$ as $n \to \infty$. A T-dimensional generalized linear model or multivariate generalized linear model is given if

1. the conditional density $f(\boldsymbol{y}_i||\boldsymbol{\vartheta}_i) = f_{\boldsymbol{y}_i|\boldsymbol{X}_i}(\boldsymbol{y}_i||\boldsymbol{\vartheta}_i)$ follows a simple T-dimensional linear exponential family with natural parameter $\boldsymbol{\vartheta}_i$, and
2. the conditional mean $\boldsymbol{\mu}_i = \mathbb{E}(\boldsymbol{y}_i|\boldsymbol{X}_i)$ of $\boldsymbol{y}_i$ given $\boldsymbol{X}_i$ is connected to the linear predictor through a one to one and sufficiently often continuously differentiable link function $\boldsymbol{g}: \mathbb{R}^T \to \mathbb{R}^T$: $\boldsymbol{g}(\boldsymbol{\mu}_i) = \boldsymbol{\eta}_i = \boldsymbol{X}_i\boldsymbol{\beta}$, where g_j might differ from $g_{j'}$ for $j, j' \in 1, \ldots, T$.

The link function $\boldsymbol{g}$ is termed natural link function, if $\boldsymbol{g}(\boldsymbol{\mu}_i) = \boldsymbol{\eta}_i = \boldsymbol{\vartheta}_i$ for $i = 1, \ldots, n$.

In analogy to the univariate case, $\boldsymbol{\mu}_i = \boldsymbol{\mu}_i(\boldsymbol{\eta}_i) = \boldsymbol{\mu}_i(\boldsymbol{X}_i, \boldsymbol{\beta}) = \boldsymbol{g}^{-1}(\boldsymbol{\eta}_i)$ is a function of $\boldsymbol{\beta}$. Furthermore, each component g_j of $\boldsymbol{g}$ is related to one and only one vector of explanatory variables $\boldsymbol{x}_{ij}$ and has the same parameter vector $\boldsymbol{\beta}$. Thus, $g_j(\boldsymbol{\mu}_i) = \boldsymbol{x}_{ij}'\boldsymbol{\beta}$, where $\boldsymbol{x}_{ij}$ is the jth column of $\boldsymbol{X}_i'$. In practice, a parameterization is preferred, where each component j has a separate parameter vector $\boldsymbol{\beta}_j$, but components have the same regression vector $\boldsymbol{x}_i$ such that $g_j(\boldsymbol{\mu}_i) = \boldsymbol{x}_i'\boldsymbol{\beta}_j$. Finally, if different components g_j and $g_{j'}$ are chosen, the entire function $\boldsymbol{g}$ needs to be considered component-wise. Thus, it is not a "general" function in this case.

3.2.2 Examples

Example 3.6 (Normal distribution — multivariate regression). For all n individuals $i = 1, \ldots, n$ let $\boldsymbol{x}_i$ be a $p \times 1$ vector of fixed and/or stochastic independent variables. Furthermore, let the T dimensional dependent variable $\boldsymbol{y}_i$ given $\boldsymbol{x}_i$ be T dimensionally normally distributed, specifically, $\boldsymbol{y}_i|\boldsymbol{x}_i \sim N_T(\boldsymbol{\mu}_i, \boldsymbol{\Sigma})$. If the identity function, i.e., $\boldsymbol{g} = \mathbf{id}$, is chosen as the link function and if $\boldsymbol{B} = (\boldsymbol{\beta}_1, \ldots, \boldsymbol{\beta}_T) \in \mathbb{R}^{p \times T}$ is the matrix comprising the parameters of interest, one obtains the multivariate linear regression model by letting $\boldsymbol{\mu}_i = \boldsymbol{\eta}_i = \boldsymbol{B}'\boldsymbol{x}_i$. This model can also be formulated in standard notation by letting $\boldsymbol{X}_i = \boldsymbol{x}_i' \otimes I_T$, which is of dimension $T \times Tp$, and the $Tp \times 1$ parameter vector $\boldsymbol{\beta} = (\boldsymbol{\beta}_1', \ldots, \boldsymbol{\beta}_p')'$. Here, $\otimes$ denotes the Kronecker product, and I_p is the $p \times p$ identity matrix.

Example 3.7 (Multinomial distribution — logistic regression). Consider n individuals and assume that the dependent variable of subject i given the covariates $\boldsymbol{X}_i$ follows a T-dimensional multinomial distribution $Mu_T(1, \boldsymbol{\pi}_i)$ for all $i = 1, \ldots, n$. Let $\boldsymbol{e}_t = (0, \ldots, 0, 1, 0 \ldots, 0)'$ denote the tth T-dimensional unit vector. Then, we have for all $\boldsymbol{y}_i = (y_{i1}, \ldots, y_{iT})'$

$$\begin{aligned} \mathbb{P}(\boldsymbol{y}_i = \boldsymbol{e}_t|\boldsymbol{X}_i) &= \pi_{it}\,, \quad \text{for} \quad t = 1, \ldots, T\,, \\ \mathbb{P}(y_{i,T+1} = 1|\boldsymbol{X}_i) &= 1 - \sum_{t=1}^{T} \pi_{it}\,, \quad \text{and} \\ \boldsymbol{\mu}_i &= \mathbb{E}(\boldsymbol{y}_i|\boldsymbol{X}_i) = \boldsymbol{\pi}_i\,. \end{aligned}$$

The linear predictor $\tilde{\boldsymbol{\eta}}_{it}$ is given by $\boldsymbol{g}_t(\boldsymbol{\mu}_i) = \tilde{\boldsymbol{\eta}}_{it} = \beta_{0t} + x_{i1}\beta_1 + \ldots + x_{ir}\beta_r$ so that $\boldsymbol{\beta}$ and $\boldsymbol{x}_{it}$ are defined as

$$\begin{aligned} \boldsymbol{\beta} &= (\beta_{01}, \ldots, \beta_{0,t-1}, \beta_{0t}, \ldots, \beta_{0T}, \beta_1, \ldots, \beta_r)' \quad \text{and} \\ \boldsymbol{x}_{it} &= (0, \ldots, 0, 1, 0, \ldots, x_{i1}, \ldots, x_{ir})'\,, \end{aligned}$$

respectively. Finally, using the natural link function $\boldsymbol{g}(\boldsymbol{\pi}_i) = \boldsymbol{\vartheta}_i$ one obtains the logistic regression model for the multinomial distribution

$$\pi_{it} = \frac{\exp(\boldsymbol{x}_{it}'\boldsymbol{\beta})}{1 + \sum_{t=1}^{T} \exp(\boldsymbol{x}_{it}'\boldsymbol{\beta})}\,.$$

An in-depth discussion of this model can be found in the literature (see, e.g., Arminger, 1995).

Example 3.8 (Multinomial distribution — cumulative logistic regression). The cumulative logistic model is an extremely popular approach for the analysis of ordered categorical data. It has been proposed by Snell (1964) and extended in several ways, (see, e.g., Fahrmeir and Tutz, 2001, pp. 75). Let the ordered categorical response of all individuals be coded as $1, \ldots, C$ so that

$C-1$ variables are needed for a complete description of the categories. The ordered categorical response z_i of subject i is extended to a response vector $\boldsymbol{y}_i$ of length $C-1$ according to

$$y_{ic} = \begin{cases} 1, & \text{if} \quad z_i \leq c\,, \\ 0, & \text{otherwise}\,. \end{cases}$$

The cumulative logistic model can then be derived as follows: Assume that the ordered categorical response z is connected through a threshold relation to an unobservable continuous stochastic variable $z^\star$:

$$z = c \quad \Longleftrightarrow \quad \vartheta_c < z^\star \leq \vartheta_c\,, \quad c = 1, \ldots, C\,,$$

where $-\infty = \vartheta_0 < \vartheta_1 < \ldots < \vartheta_C = +\infty$. Furthermore, connect the latent variable $z^\star$ with the regressor variables $\boldsymbol{x}$ for all individuals i by the linear model

$$z^\star = -\boldsymbol{x}'\tilde{\boldsymbol{\beta}} + \epsilon^\star\,,$$

where $\tilde{\boldsymbol{\beta}}$, as before, is the vector of regression parameters, and $\epsilon^\star$ is a latent random error variable with distribution function F.

Then, the observable variable z is determined by

$$\mathbb{P}(z \leq c|\boldsymbol{x}) = F(\vartheta_c + \boldsymbol{x}'\tilde{\boldsymbol{\beta}})\,. \tag{3.2}$$

This model is termed a cumulative model with distribution function F, because the left side of Eq. 3.2 is a sum of probabilities. The choice of the logistic function yields the cumulative logistic regression model

$$\mathbb{P}(z \leq c|\boldsymbol{x}) = \frac{\exp(\vartheta_c + \boldsymbol{x}'\tilde{\boldsymbol{\beta}})}{1 + \exp(\vartheta_c + \boldsymbol{x}'\tilde{\boldsymbol{\beta}})} \tag{3.3}$$

for $c = 1, \ldots, C-1$. Equation 3.3 can equivalently be written as

$$\ln \left\{ \frac{\mathbb{P}(z \leq c|\boldsymbol{x})}{\mathbb{P}(z > c|\boldsymbol{x})} \right\} = \vartheta_c + \boldsymbol{x}'\tilde{\boldsymbol{\beta}}\,,$$

and it is seen that the regression lines $\boldsymbol{x}'\tilde{\boldsymbol{\beta}}$ are parallel. The model is therefore also termed the proportional odds model.

If the model includes a regression constant, $C-2$ dummy variables are needed for threshold modeling. This can be seen easily from the matrix notation of the model:

$$\boldsymbol{X}_i = \begin{pmatrix} 1 & & & & \boldsymbol{x}_i' \\ & 1 & & & \boldsymbol{x}_i' \\ & & \ddots & & \\ & & & 1 & \boldsymbol{x}_i' \\ 0\,0 & 0 & & 0 & \boldsymbol{x}_i' \end{pmatrix}, \quad \text{and} \quad \boldsymbol{\beta} = (\vartheta_1, \ldots, \vartheta_{C-2}, \tilde{\boldsymbol{\beta}}')' .$$

The matrix $\boldsymbol{X}_i$ thus is of dimension $(C-1) \times (C-2+p)$.

Chapter 4
Maximum likelihood method

The most popular estimation approach is the maximum likelihood (ML) method. In this chapter, the ML estimator is defined first, and important asymptotic properties of the ML estimator are formulated in Sect. 4.2. Transformations of estimators, not only ML estimators, are discussed in Sect. 4.3. To illustrate the ML approach, we consider the ML method in the linear exponential family (Sect. 4.4) and in univariate GLM (Sect. 4.5). A crucial assumption of ML estimation is the correct specification of the underlying statistical model. Therefore, we discuss the consequences of using the ML method in misspecified models in Sect. 4.6. Even if the model is misspecified, it is based on a likelihood, and the resulting estimator is therefore called a quasi maximum likelihood (QML) estimator (for an in-depth discussion, see White, 1982, 1994). The reader should note that QML estimation is different from quasi likelihood (QL) estimation. The latter approach is a generalization of the generalized linear model (McCullagh and Nelder, 1989; Wedderburn, 1974) and requires the correct specification of the first two moments.

4.1 Definition

To define an ML estimator, we do not require many assumptions. However, important properties of the ML estimator can be derived only under specific regularity assumptions. These aspects are discussed in some detail below. We start with some notation.

Let $\boldsymbol{y}_i$ be a $T \times 1$ stochastic vector and $\boldsymbol{X}_i$ a $T \times p$ stochastic and/or fixed matrix. The pairs $(\boldsymbol{y}_i, \boldsymbol{X}_i)$ are assumed to be independent, and $\boldsymbol{y}_i|\boldsymbol{X}_i$ are assumed to be identically distributed for all $i = 1, \ldots, n$. The true conditional density (or probability mass function for discrete random vectors) $f^\star(\boldsymbol{y}_i|\boldsymbol{X}_i||\boldsymbol{\beta})$ of $\boldsymbol{y}_i$ given $\boldsymbol{X}_i$ depends on a parameter vector $\boldsymbol{\beta} \in \Theta \subset \mathbb{R}^p$. Furthermore, we assume for both continuous and discrete distributions that $f^\star$ is correctly specified. The aim is to estimate the unknown p-dimensional parameter vector $\boldsymbol{\beta}$. Before we can introduce the ML estimator, we have to consider the joint distribution of $\boldsymbol{y}_i$ and $\boldsymbol{X}_i$. Specifically, we assume that the marginal density $m(\boldsymbol{X}_i)$ is independent of $\boldsymbol{\beta}$. As a result, the joint density f of $\boldsymbol{y}_i$ and $\boldsymbol{X}_i$ is given by

$$f(\boldsymbol{y}_i, \boldsymbol{X}_i || \boldsymbol{\beta}) = f^\star(\boldsymbol{y}_i | \boldsymbol{X}_i || \boldsymbol{\beta}) m(\boldsymbol{X}_i)\,.$$

For probability calculations, we assume that all parameters are known and that the data $\boldsymbol{y}_i$ are unknown, i.e., unobserved. For parameter estimation, we pretend as if the observations were known. Subsequently, we consider the individual likelihood function $L_i(\boldsymbol{\beta} || \boldsymbol{y}_i, \boldsymbol{X}_i) = L_i(\boldsymbol{\beta})$ for $\boldsymbol{\beta}$ given the data $(\boldsymbol{y}_i, \boldsymbol{X}_i)$. Because we assume that the data are given, we do distinguish between random vectors and their realizations.

The pairs $(\boldsymbol{y}_i, \boldsymbol{X}_i)$ are assumed to be independent so that the (global) likelihood function, i.e., the joint density of all clusters, is the product of the individual likelihood functions $L(\boldsymbol{\beta}) = \prod_{i=1}^n L_i(\boldsymbol{\beta})$.

Definition 4.1 (Maximum likelihood estimator). A maximum likelihood estimator (MLE) of $\boldsymbol{\beta}$ is a solution to the maximization problem

$$\max_{\boldsymbol{\beta} \in \Theta \subset \mathrm{IR}^p} L(\boldsymbol{\beta} || \boldsymbol{y}_i, \boldsymbol{X}_i)\,.$$

In many applications, the logarithm of the likelihood function is considered, and

$$\tilde{l}(\boldsymbol{\beta}) = \frac{1}{n} \ln L(\boldsymbol{\beta}) = \frac{1}{n} \sum_{i=1}^n \ln L_i(\boldsymbol{\beta}) = \frac{1}{n} \sum_{i=1}^n \Big(\ln f^\star(\boldsymbol{y}_i | \boldsymbol{X}_i || \boldsymbol{\beta}) + \ln m(\boldsymbol{X}_i) \Big)$$

is called the normed loglikelihood function. The logarithm is a strictly isotone function so that the solution to the maximization problem is not altered.

The maximization problem using the normed loglikelihood function is independent of the marginal density m, and m is irrelevant for the maximization. One therefore considers the kernel of the normed loglikelihood function for maximization. The kernel only contains the parts of the normed loglikelihood function that are relevant for maximization, and the kernel of the normed likelihood function is therefore given by

$$l(\boldsymbol{\beta}) = \frac{1}{n} \sum_{i=1}^n l_i(\boldsymbol{\beta}) = \frac{1}{n} \sum_{i=1}^n \ln f^\star(\boldsymbol{y}_i | \boldsymbol{X}_i || \boldsymbol{\beta})\,. \tag{4.1}$$

For this reason, the maximization problem for stochastic and fixed regressors is identical. However, the asymptotic properties are different; for details see, e.g., Greene (2007).

One problem related to ML estimation is that a modification of the density on a set of points having zero probability may alter the resulting estimator (see, e.g., Gourieroux and Monfort, 1995a). Therefore, whenever possible, one assumes that densities are continuous in $\boldsymbol{y}_i$ or, at least, piecewise continuous. The second aspect is related to the existence of parameter estimates. In fact, nonexistence may occur if the parameter space Θ is open or if the loglikelihood function is not continuous. Therefore, to guarantee existence of parameter estimates, one assumes that the parameter space Θ is compact and that the likelihood function is continuous on Θ. These assumptions have been summarized, e.g., by White (1982, assumptions A1 and A2). Finally, even if there is a maximum likelihood estimator, it need not be unique. A sufficient condition for its uniqueness is the strict concavity of the likelihood function in a bijective transformation of the parameter $\boldsymbol{\beta}$. Because the uniqueness cannot be guaranteed, one generally tries to find local maxima of the kernel of the normed likelihood function from Eq. 4.1.

Often, likelihood functions are considered that are at least two times continuously differentiable. ML estimators are then obtained by differentiating the kernel of the normed likelihood function with respect to $\boldsymbol{\beta}$, i.e.,

$$\boldsymbol{u}(\boldsymbol{\beta}) = \frac{\partial l(\boldsymbol{\beta})}{\partial \boldsymbol{\beta}} = \frac{1}{n}\sum_{i=1}^{n} \boldsymbol{u}_i(\boldsymbol{\beta}) = \frac{1}{n}\sum_{i=1}^{n} \frac{\partial l_i(\boldsymbol{\beta})}{\partial \boldsymbol{\beta}} . \tag{4.2}$$

An ML estimator $\hat{\boldsymbol{\beta}}$ is the root of Eq. 4.2, which is termed the score function. Thus, one aims to find the solution of $\boldsymbol{u}(\hat{\boldsymbol{\beta}}) = \mathbf{0}$, and this equation is called the maximum likelihood equation (MLE).

To obtain local maxima, the Hessian matrix, i.e., the matrix of second derivatives,

$$\boldsymbol{W}(\boldsymbol{\beta}) = \frac{\partial^2 l(\boldsymbol{\beta})}{\partial \boldsymbol{\beta} \partial \boldsymbol{\beta}'} = \frac{1}{n}\sum_{i=1}^{n} \boldsymbol{W}_i(\boldsymbol{\beta}) = \frac{1}{n}\sum_{i=1}^{n} \frac{\partial^2 l_i(\boldsymbol{\beta})}{\partial \boldsymbol{\beta} \partial \boldsymbol{\beta}'} , \tag{4.3}$$

has to be negative definite.

4.2 Asymptotic properties

We assume that the usual maximum likelihood regularity conditions A1–A7 as given, e.g., by White (1982), are fulfilled. They can be sketched as follows (for the exact formulations, see White, 1982):

- The independent random vectors $\boldsymbol{y}_i$ have a distribution with some Radon-Nikodym density (termed g by White), and the parametric family of distribution functions all have densities $f(\boldsymbol{y}||\boldsymbol{\beta})$.
- $f(\boldsymbol{y}||\boldsymbol{\beta})$ and $\partial \ln f / \partial \boldsymbol{\beta}$ are measurable in $\boldsymbol{y}$ and continuous in $\boldsymbol{\beta}$.

- $\partial^2 \ln f/\partial\beta_j\beta_{j'}$, $\partial \ln f/\partial\beta_j \cdot \partial \ln f/\partial\beta_{j'}$ and $\partial(\partial f/\partial\beta_j \cdot f)/\partial\beta_{j'}$ are dominated by functions integrable in $\boldsymbol{\beta}$.
- The parameter space is compact.
- $\mathbb{E}|\ln g| < \infty$, $|\ln f|$ is bounded uniformly in $\boldsymbol{\beta}$.
- The Kullback-Leibler information criterion (for the definition, see Eq. 4.21) $\mathbb{E}\ln\big(g/f\big)$ has a unique maximum at the true parameter value $\boldsymbol{\beta}_0$.
- $\boldsymbol{\beta}_0$ is in the interior of the parameter space, the outer product gradient is regular (for the definition, see Theorem 4.2), and the rank of the Fisher information matrix (for the definition, see Theorem 4.2) is constant in a neighborhood of $\boldsymbol{\beta}_0$.
- The minimal support of $f(\boldsymbol{y}||\boldsymbol{\beta})$ does not depend on $\boldsymbol{\beta}$.

Then, one can show the following statements.

Theorem 4.2 (Properties of ML estimators).

1. *There asymptotically exists an ML estimator $\hat{\boldsymbol{\beta}}$ for the true parameter vector $\boldsymbol{\beta}_0$.*
2. *The ML estimator $\hat{\boldsymbol{\beta}}$ converges almost surely to the true parameter $\boldsymbol{\beta}_0$.*
3. *The ML estimator $\hat{\boldsymbol{\beta}}$ for $\boldsymbol{\beta}_0$ is asymptotically normal. More specifically, with $\stackrel{a}{\sim}$ denoting "asymptotically distributed as," we get*
$$\sqrt{n}(\hat{\boldsymbol{\beta}} - \boldsymbol{\beta}_0) \stackrel{a}{\sim} N\big(\mathbf{0}, \boldsymbol{B}(\boldsymbol{\beta}_0)^{-1}\big)\,, \tag{4.4}$$
where $\boldsymbol{B}(\boldsymbol{\beta}) = \mathbb{E}^{\boldsymbol{X}}\big(\mathbb{E}^{\boldsymbol{y}} - \boldsymbol{u}_i(\boldsymbol{\beta})\boldsymbol{u}_i(\boldsymbol{\beta})'\big)$ is the outer product of the score vector and therefore termed the outer product gradient (OPG). It is also termed the outer product of the Fisher information matrix.
4. *Because the likelihood is assumed to be correctly specified, the OPG equals the Fisher information matrix. Thus, it is equal to the expectation of the negative Hessian matrix of subject i: $\boldsymbol{A}(\boldsymbol{\beta}) = \mathbb{E}^{\boldsymbol{X}}\big(\mathbb{E}^{\boldsymbol{y}} - \boldsymbol{W}_i(\boldsymbol{\beta})\big)$. Therefore, we also have*
$$\sqrt{n}(\hat{\boldsymbol{\beta}} - \boldsymbol{\beta}_0) \stackrel{a}{\sim} N\big(\mathbf{0}, \boldsymbol{A}(\boldsymbol{\beta}_0)^{-1}\big)\,.$$
5. *A (strongly) consistent estimator of the Fisher information matrix $\boldsymbol{A}(\boldsymbol{\beta}_0)$ is, e.g., given by $\hat{\boldsymbol{A}}(\hat{\boldsymbol{\beta}})$, which is the Fisher information matrix evaluated at $\hat{\boldsymbol{\beta}}$. $\hat{\boldsymbol{A}}(\hat{\boldsymbol{\beta}})$ is termed the observed Fisher information matrix. Alternative (strongly) consistent estimators for $\boldsymbol{A}(\boldsymbol{\beta}_0)$ include*
$$-\hat{\boldsymbol{W}}(\hat{\boldsymbol{\beta}}) = -\frac{1}{n}\frac{\partial^2 l(\hat{\boldsymbol{\beta}})}{\partial\boldsymbol{\beta}\partial\boldsymbol{\beta}'} \quad \textit{and} \quad -\frac{1}{n}\sum_{i=1}^{n}\frac{\partial^2 l_i(\hat{\boldsymbol{\beta}})}{\partial\boldsymbol{\beta}\partial\boldsymbol{\beta}'}\,.$$
6. *Strongly consistent estimators $\hat{\boldsymbol{B}}(\hat{\boldsymbol{\beta}})$ of the OPG are, e.g., given $Bf(\boldsymbol{\beta}_0)$ by*

$$\frac{1}{n}\frac{\partial l(\hat{\boldsymbol{\beta}})}{\partial \boldsymbol{\beta}}\frac{\partial l(\hat{\boldsymbol{\beta}})}{\partial \boldsymbol{\beta}'}\,, \quad \text{or by} \quad \frac{1}{n}\sum_{i=1}^{n}\hat{\boldsymbol{u}}_i(\hat{\boldsymbol{\beta}})\hat{\boldsymbol{u}}_i(\hat{\boldsymbol{\beta}})' = \frac{1}{n}\sum_{i=1}^{n}\frac{\partial l_i(\hat{\boldsymbol{\beta}})}{\partial \boldsymbol{\beta}}\frac{\partial l_i(\hat{\boldsymbol{\beta}})}{\partial \boldsymbol{\beta}'}\,.$$

$\hat{\boldsymbol{B}}(\hat{\boldsymbol{\beta}})$ is the termed estimated OPG or estimated outer product of the Fisher information matrix.

7. *The ML estimator $\hat{\boldsymbol{\beta}}$ is asymptotically efficient and thus reaches the Rao–Cramér bound (Rao, 1973, p. 350).*

Before we prove this theorem, we make several remarks.

Remark 4.3.

- An estimator with Properties 1., 2., 3., and 7. of the theorem is called a best asymptotically normally (BAN) distributed estimator.
- The OPG and the Fisher information matrix need not be equal; for an example, see Sect. 4.6.1.
- If a nuisance parameter $\boldsymbol{\Psi}$ is added, such as the variance matrix $\boldsymbol{\Sigma}$ for the normal distribution, the results of Theorem 4.2 can be extended in the following sense: One replaces the nuisance parameter $\boldsymbol{\Psi}$ by a (strongly) $\sqrt{n}$ consistent estimator $\hat{\boldsymbol{\Psi}}$ for the true nuisance parameter $\boldsymbol{\Psi}_0$. Note that $\hat{\boldsymbol{\Psi}}$ is strongly $\sqrt{n}$ consistent if $\sqrt{n}(\hat{\boldsymbol{\Psi}} - \boldsymbol{\Psi})$ is bounded with probability 1. The estimator $\hat{\boldsymbol{\beta}}$ maximizing the normed loglikelihood function $\frac{1}{n}\sum_{i=1}^{n} \ln f(\boldsymbol{y}_i, \boldsymbol{X}_i || \boldsymbol{\beta}, \hat{\boldsymbol{\Psi}})$ has the same properties as the estimator $\hat{\boldsymbol{\beta}}$ of Theorem 4.2 (see, e.g., Gourieroux et al., 1984b, p. 682).
- The results given of Theorem 4.2 are large sample results only. To give an example, the ML estimator is asymptotically unbiased but it may be biased in finite samples. A simple example is the ML estimator of the variance σ^2 in the case of a normal distribution with unknown mean μ. The ML estimator is $\hat{\sigma}^2 = \frac{1}{n}\sum_{i=1}^{n}\sum_{i=1}^{n}(x_i - x)^2$, which has expected value $\frac{n-1}{n}\sigma^2$.

Proof. Here, we sketch the proof of statement 3. and show statement 4. because the result on the asymptotic distribution will be used in subsequent chapters. References are given for the other statements.
1.: See, e.g., White (1982, Theorem 2.1).
2.: See, e.g., White (1982, Theorem 2.2).
5.: See, e.g., White (1982, Theorem 3.2); also see, e.g., Gourieroux and Monfort (1995a, p. 186).
6.: See, e.g., White (1982, Theorem 3.3).
7.: See, e.g., Gourieroux and Monfort (1995a, p. 184).
3.: For simplicity, we drop the index i. The derivatives $\boldsymbol{u}_i(\boldsymbol{\beta})$ and $\boldsymbol{W}_i(\boldsymbol{\beta})$ are denoted by $\boldsymbol{s}(\boldsymbol{\beta})$ and $\boldsymbol{S}(\boldsymbol{\beta})$, respectively.

In the first step, the ML estimator of Eq. 4.2 is approximated by a first-order Taylor expansion around $\boldsymbol{\beta}_0$:

$$\mathbf{0} \overset{a.s.}{=} \boldsymbol{u}(\boldsymbol{\beta}_0) + \boldsymbol{W}(\boldsymbol{\beta}^*)(\hat{\boldsymbol{\beta}} - \boldsymbol{\beta}_0) = \frac{1}{n}\sum_{i=1}^{n} \boldsymbol{u}_i(\boldsymbol{\beta}_0) + \Big(\frac{1}{n}\sum_{i=1}^{n} \boldsymbol{W}_i(\boldsymbol{\beta}^*)\Big)(\hat{\boldsymbol{\beta}} - \boldsymbol{\beta}_0),$$

with a.s. denoting almost surely, and $\boldsymbol{\beta}^*$ lying on the line segment between $\hat{\boldsymbol{\beta}}$ and $\boldsymbol{\beta}_0$, i.e., $|\boldsymbol{\beta}^* - \boldsymbol{\beta}_0| \leq |\hat{\boldsymbol{\beta}} - \boldsymbol{\beta}_0|$.

Application of a theorem by Cramér-Slutsky (see, e.g., Rohatgi and Saleh, 2001, p. 270, Theorem 15 (c)) and pre-multiplication by $\sqrt{n}$ yields

$$\sqrt{n}(\hat{\boldsymbol{\beta}} - \boldsymbol{\beta}_0) \overset{a.s.}{=} \Big(-\frac{1}{n}\sum_{i=1}^{n} \boldsymbol{W}_i(\boldsymbol{\beta}^*)\Big)^{-1} \Big(\frac{1}{\sqrt{n}}\sum_{i=1}^{n} \boldsymbol{u}_i(\boldsymbol{\beta}_0)\Big). \tag{4.5}$$

By a strong law of large numbers (White, 1981, Lemma 3.1), $-\frac{1}{n}\sum_{i=1}^{n} \boldsymbol{W}_i(\boldsymbol{\beta}^*)$ converges to the Fisher information matrix $\mathbb{E}^{\boldsymbol{X}}\left(\mathbb{E}^{\boldsymbol{y}} \boldsymbol{S}(\boldsymbol{\beta}_0)\right) = \boldsymbol{A}(\boldsymbol{\beta}_0)$.

According to the regularity conditions, differentiation and integration may be interchanged so that the expectation of the score vector is $\mathbf{0}$:

$$\begin{aligned}
\mathbb{E}^{\boldsymbol{X}}\mathbb{E}^{\boldsymbol{y}}\left(\frac{1}{\sqrt{n}}\sum_{i=1}^{n} \boldsymbol{u}_i(\boldsymbol{\beta}_0)\right) &= \sqrt{n}\,\mathbb{E}^{\boldsymbol{X}}\mathbb{E}^{\boldsymbol{y}}\left(\boldsymbol{u}(\boldsymbol{\beta}_0)\right) \\
&= \sqrt{n}\,\mathbb{E}^{\boldsymbol{X}}\Big(\int \frac{\partial \ln f(\boldsymbol{y}|\boldsymbol{X}||\boldsymbol{\beta}_0)}{\partial \boldsymbol{\beta}} f(\boldsymbol{y}|\boldsymbol{X}||\boldsymbol{\beta}_0)\, \boldsymbol{dy}\Big) \\
&= \sqrt{n}\,\mathbb{E}^{\boldsymbol{X}}\Big(\frac{\partial}{\partial \boldsymbol{\beta}} \int f(\boldsymbol{y}|\boldsymbol{X}||\boldsymbol{\beta}_0)\, \boldsymbol{dy}\Big) = \mathbf{0}.
\end{aligned}$$

Here, $\partial \ln f(\boldsymbol{y}|\boldsymbol{X}, \boldsymbol{\beta}_0)/\partial \boldsymbol{\beta}$ denotes the evaluation of the first derivative of $\ln f(\boldsymbol{y}|\boldsymbol{X}, \boldsymbol{\beta}_0)$ at $\boldsymbol{\beta}_0$.

The covariance matrix of the score vector can be obtained by using the i.i.d. assumption and $\mathbb{E}(\boldsymbol{u}) = \mathbf{0}$:

$$\mathbb{V}\mathrm{ar}\Big(\frac{1}{\sqrt{n}}\sum_{i=1}^{n} \boldsymbol{u}_i(\boldsymbol{\beta}_0)\Big) = \frac{1}{n}\sum_{i=1}^{n} \mathbb{V}\mathrm{ar}(\boldsymbol{u}_i(\boldsymbol{\beta}_0)) = \mathbb{E}^{\boldsymbol{X}}\mathbb{E}^{\boldsymbol{y}}\Big(\boldsymbol{s}(\boldsymbol{\beta}_0)\boldsymbol{s}(\boldsymbol{\beta}_0)'\Big) = \boldsymbol{B}(\boldsymbol{\beta}_0).$$

The asymptotic distribution is obtained by using the multivariate central limit theorem (see, e.g., Lehmann and Casella, 1998, p. 61, Theorem 8.21)

$$\frac{1}{\sqrt{n}}\sum_{i=1}^{n} \boldsymbol{u}_i(\boldsymbol{\beta}_0) \overset{a}{\sim} N\left((\mathbf{0}, \boldsymbol{B}(\boldsymbol{\beta}_0)\right). \tag{4.6}$$

Equation 4.5, Cramér-Slutsky's theorem (see, e.g., Rohatgi and Saleh, 2001, p. 269, Theorem 14 together with p. 270, Theorem 15 (c)), and the convergence of the Hessian matrix to the Fisher information matrix give the asymptotic distribution of $\sqrt{n}(\hat{\boldsymbol{\beta}} - \boldsymbol{\beta}_0)$:

$$\sqrt{n}(\hat{\boldsymbol{\beta}} - \boldsymbol{\beta}_0) \overset{a}{\sim} N\left(\mathbf{0}, [\boldsymbol{A}(\boldsymbol{\beta}_0)]^{-1}\boldsymbol{B}(\boldsymbol{\beta}_0)[\boldsymbol{A}(\boldsymbol{\beta}_0)]^{-1}\right). \tag{4.7}$$

This formulation of the asymptotic distribution of $\sqrt{n}(\hat{\boldsymbol{\beta}}-\boldsymbol{\beta}_0)$ involves the OPG as well as the Fisher information matrix. The matrix $\boldsymbol{C} = \boldsymbol{A}^{-1}\boldsymbol{B}\boldsymbol{A}^{-1}$ is termed the "sandwich matrix" with $-\boldsymbol{A}$ being the "bread" and $\boldsymbol{B}$ being the "butter." It is also named the robust covariance matrix. The interpretation of the latter term is discussed in detail in the next chapter.

4.: Statement 4. remains to be shown, i.e., $\boldsymbol{A}(\boldsymbol{\beta}_0) = \boldsymbol{B}(\boldsymbol{\beta}_0)$ for completion of statement 3. This equality of the OPG and the Fisher information matrix can be shown under the assumption that the likelihood function is correctly specified.

For simplicity, we omit integration over $\boldsymbol{X}$ in the following. By using the chain rule and the differentiation rules for quotients and logarithms, we get

$$\frac{\partial^2 \ln f(\boldsymbol{y}|\boldsymbol{X}||\boldsymbol{\beta}_0)}{\partial\boldsymbol{\beta}\partial\boldsymbol{\beta}'} = \frac{\dfrac{\partial^2 f(\boldsymbol{y}|\boldsymbol{X}||\boldsymbol{\beta}_0)}{\partial\boldsymbol{\beta}\partial\boldsymbol{\beta}'}}{f(\boldsymbol{y}|\boldsymbol{X}||\boldsymbol{\beta}_0)} - \frac{\left(\dfrac{\partial f(\boldsymbol{y}|\boldsymbol{X}||\boldsymbol{\beta}_0)}{\partial\boldsymbol{\beta}}\right)\left(\dfrac{\partial f(\boldsymbol{y}|\boldsymbol{X}||\boldsymbol{\beta}_0)}{\partial\boldsymbol{\beta}}\right)'}{f(\boldsymbol{y}|\boldsymbol{X}||\boldsymbol{\beta}_0)^2}\,.$$

The equality $\boldsymbol{A} = \boldsymbol{B}$ can now be shown easily:

$$\begin{aligned}
\mathbb{E}^{\boldsymbol{y}}\boldsymbol{S}(\boldsymbol{\beta}_0) &= \int \frac{\partial^2 \ln f(\boldsymbol{y}|\boldsymbol{X}||\boldsymbol{\beta}_0)}{\partial\boldsymbol{\beta}\partial\boldsymbol{\beta}'} f(\boldsymbol{y}|\boldsymbol{X}||\boldsymbol{\beta}_0)\,\boldsymbol{dy} \\
&= \frac{\partial^2}{\partial\boldsymbol{\beta}\partial\boldsymbol{\beta}'}\int f(\boldsymbol{y}|\boldsymbol{X}||\boldsymbol{\beta}_0)\,\boldsymbol{dy} \\
&\quad -\int \left(\frac{\partial \ln f(\boldsymbol{y}|\boldsymbol{X}||\boldsymbol{\beta}_0)}{\partial\boldsymbol{\beta}}\right)\left(\frac{\partial \ln f(\boldsymbol{y}|\boldsymbol{X}||\boldsymbol{\beta}_0)}{\partial\boldsymbol{\beta}}\right)' f(\boldsymbol{y}|\boldsymbol{X}||\boldsymbol{\beta}_0)\,\boldsymbol{dy} \\
&= -\mathbb{E}^{\boldsymbol{y}}\left(\boldsymbol{s}(\boldsymbol{\beta}_0)\boldsymbol{s}(\boldsymbol{\beta}_0)'\right).
\end{aligned}$$

In summary, Eq. 4.7 reduces to the simple form

$$\sqrt{n}(\hat{\boldsymbol{\beta}}-\boldsymbol{\beta}_0) \overset{a}{\sim} N\left(\boldsymbol{0},[\boldsymbol{A}(\boldsymbol{\beta}_0)]^{-1}\right) = N\left(\boldsymbol{0},[\boldsymbol{B}(\boldsymbol{\beta}_0)]^{-1}\right)\,, \tag{4.8}$$

and the proof of both 3. and 4. is completed. □

4.3 Transformations

In this section, we investigate the behavior of ML estimators under transformations. First, we consider the ML estimator of a bijective transformation.

Theorem 4.4 (Invariance principle for ML estimators). *Consider a likelihood function with parameter vector $\boldsymbol{\beta}$ and a bijective function $\boldsymbol{v}$ from Θ on to a set $\boldsymbol{\Lambda}$. If $\hat{\boldsymbol{\beta}}$ is the ML estimator of $\boldsymbol{\beta}$, then $\hat{\boldsymbol{\xi}} = \boldsymbol{v}(\hat{\boldsymbol{\beta}})$ is the ML estimator of $\boldsymbol{\xi} \in \Xi = \boldsymbol{v}(\Theta)$ – corresponding to the likelihood function defined on Ξ.*

Proof. See, e.g., Gourieroux and Monfort (1995a, p. 175). □

Remark 4.5. The result formulated in Theorem 4.4 is important for applications. For example, we are often interested in estimating the standard deviation of a normally distributed random variable. In this case, an estimator of the variance can be derived easily using the ML method. However, an estimator for the standard deviation is not ready at hand. Therefore, it is good to know that one only needs to take the square root of the ML estimator of the variance to obtain the ML estimator of the standard deviation.

The next theorem states that a transformed estimator is asymptotically normally distributed if the original estimator is asymptotically normal and if the transformation function is continuously differentiable.

Theorem 4.6 (Multivariate delta method). *Consider an estimator $\hat{\boldsymbol{\beta}}$ for $\boldsymbol{\beta}_0$ that is asymptotically normally distributed, in detail, $\hat{\boldsymbol{\beta}} \overset{a}{\sim} N(\boldsymbol{\beta}_0, \mathbb{V}ar(\boldsymbol{\beta}_0))$. We assume that a transformation function $\boldsymbol{\xi} = \boldsymbol{v}(\boldsymbol{\beta})$ of $\boldsymbol{\beta}$ is continuously differentiable with respect to $\boldsymbol{\beta}$ in a neighborhood of $\boldsymbol{\beta}_0$. The estimator $\hat{\boldsymbol{\xi}} = \boldsymbol{v}(\hat{\boldsymbol{\beta}})$ of $\boldsymbol{\xi}_0 = \boldsymbol{v}(\boldsymbol{\beta}_0)$ is asymptotically normal, precisely:*

$$\hat{\boldsymbol{\xi}} \overset{a}{\sim} N\left(\boldsymbol{\xi}_0, \frac{\partial \boldsymbol{\xi}(\boldsymbol{\beta}_0)}{\partial \boldsymbol{\beta}'} \mathbb{V}ar(\boldsymbol{\beta}_0) \frac{\partial \boldsymbol{\xi}(\boldsymbol{\beta}_0)'}{\partial \boldsymbol{\beta}}\right).$$

The covariance matrix of $\hat{\boldsymbol{\xi}}$ is estimated by replacing $\boldsymbol{\beta}_0$ with $\hat{\boldsymbol{\beta}}$.

Proof. If we admit that $\boldsymbol{v}(\boldsymbol{\beta})$ can be expanded in a Taylor series around $\boldsymbol{v}(\boldsymbol{\beta}_0)$, we obtain $\sqrt{n}(\boldsymbol{v}(\hat{\boldsymbol{\beta}}) - \boldsymbol{v}(\boldsymbol{\beta}_0)) \overset{a.s.}{=} (\partial \boldsymbol{v}(\boldsymbol{\beta}_0)/\partial \boldsymbol{\beta}')\sqrt{n}(\hat{\boldsymbol{\beta}} - \boldsymbol{\beta}_0)$. The left side thus is asymptotically equivalent to a linear function of a random vector of which we know its asymptotic normal distribution, and the covariance matrix can be obtained using standard calculation rules for covariance matrices. □

The inversion of the idea of the multivariate delta method leads to the minimum distance estimation (MDE) approach. Specifically, we consider the case that $\boldsymbol{\beta} = \boldsymbol{\beta}(\boldsymbol{\kappa})$ is some function of a parameter vector $\boldsymbol{\kappa} \in K \subset \mathbb{R}^q$, $q \leq p$. Regularity conditions (for details, see Küsters, 1987) include that $\boldsymbol{\kappa}$ is first-order identifiable, i.e., $\boldsymbol{\beta}(\boldsymbol{\kappa}_1) = \boldsymbol{\beta}(\boldsymbol{\kappa}_2) \Rightarrow \boldsymbol{\kappa}_1 \overset{a.s.}{=} \boldsymbol{\kappa}_2$, and that the number of restrictions does not exceed the dimension of $\boldsymbol{\beta}$.

Definition 4.7 (Minimum distance estimator). The minimum distance estimator $\hat{\boldsymbol{\kappa}}$ of $\boldsymbol{\kappa}$ is the minimum over all $\boldsymbol{\kappa}$, precisely, the minimum of the Mahalanobis distance $Q(\boldsymbol{\kappa})$:

$$\min_{\boldsymbol{\kappa} \in K \subset \mathbb{R}^q} Q(\boldsymbol{\kappa}) = n\left(\hat{\boldsymbol{\beta}} - \boldsymbol{\beta}(\boldsymbol{\kappa})\right)' \left(\mathbb{V}\text{ar}(\hat{\boldsymbol{\beta}})\right)^{-1} \left(\hat{\boldsymbol{\beta}} - \boldsymbol{\beta}(\boldsymbol{\kappa})\right) \tag{4.9}$$

Theorem 4.8 (Minimum distance estimation). *The minimum distance estimator $\hat{\boldsymbol{\kappa}}$ of $\boldsymbol{\kappa}_0$ is asymptotically normally distributed:*

$$\sqrt{n}(\hat{\boldsymbol{\kappa}} - \boldsymbol{\kappa}_0) \overset{a}{\sim} N\left(\mathbf{0}, \left[\frac{\partial \boldsymbol{\beta}(\boldsymbol{\kappa}_0)'}{\partial \boldsymbol{\kappa}} \mathbb{V}ar\big(\sqrt{n}\hat{\boldsymbol{\beta}}\big)^{-1} \frac{\partial \boldsymbol{\beta}(\boldsymbol{\kappa}_0)}{\partial \boldsymbol{\kappa}'}\right]^{-1}\right). \tag{4.10}$$

The covariance matrix of $\hat{\boldsymbol{\kappa}}$ is estimated by replacing $\boldsymbol{\kappa}_0$ with $\hat{\boldsymbol{\kappa}}$.

Under the null hypothesis $H_0 \colon \boldsymbol{\beta} = \boldsymbol{\beta}(\boldsymbol{\kappa})$, the Mahalanobis distance of $Q(\boldsymbol{\kappa})$ of Eq. 4.9 is asymptotically χ^2 distributed with the number of free parameters being the degrees of freedom.

Proof. The proof has been given, e.g., by Arminger (1995); technical details including regularity conditions can be found in Küsters (1987). The proof is based on a first-order Taylor series from which the first statement follows. The second statement about the asymptotic χ^2 distribution is a direct consequence of the quadratic form. □

We now give several examples for parameter reparameterizations that are often used in applications.

Example 4.9 (Common parameter reparameterizations).

- Equality restrictions $\beta_i = \beta_j$ are obtained by setting $\beta_i = \kappa_k$ and $\beta_j = \kappa_k$.
- Linear restrictions of the form $\sum_{j=1}^{J} a_j \beta_j = d$ with a_j and d as known constants may be written as functions of unrestricted parameters $\kappa_1, \ldots, \kappa_{J-1}$ via $\beta_j = \kappa_j$, for $j = 1, \ldots, J-1$, and $\beta_J = \big(d - \sum_{j=1}^{J-1} \kappa_j\big)/c_J$.
- Domain restrictions of the form $\beta_j \in]a_j, b_j[$ can be eliminated by expit transformations with unrestricted κ_j: $\beta_j = a_j + (b_j - a_j)\text{expit}\big(\kappa_j\big)$.
- Inequality restrictions of the form $0 \le \beta_1 \le \beta_2 \le \ldots \le \beta_J$ can be reparameterized via $\beta_1 = \kappa_1^2$, $\beta_2 = \kappa_1^2 + \kappa_2^2$, up to $\beta_K = \sum_{j=1}^{J} \kappa_j^2$.

4.4 Maximum likelihood estimation in linear exponential families

ML estimation is a very general approach, and it is simplified substantially in linear exponential families. We consider n independently but not necessarily identically distributed T-dimensional random vectors $\boldsymbol{y}_1, \ldots, \boldsymbol{y}_n$ with densities (or probability mass functions for discrete distributions) belonging to the simple linear exponential family

$$f(\boldsymbol{y}_i||\boldsymbol{\vartheta}, \boldsymbol{\Psi}) = \exp\Big(\boldsymbol{\vartheta}'\boldsymbol{y}_i + b_i(\boldsymbol{y}_i, \boldsymbol{\Psi}) - d_i(\boldsymbol{\vartheta}, \boldsymbol{\Psi})\Big).$$

If the specific distribution includes a nuisance parameter, we assume that the nuisance parameter is either known or can be replaced by a $\sqrt{n}$ consistent estimator.

The kernel of the normed loglikelihood function is given by $l(\boldsymbol{\vartheta}) = \frac{1}{n}\sum_{i=1}^{n}\big(\boldsymbol{\vartheta}'\boldsymbol{y}_i - d_i(\boldsymbol{\vartheta}, \boldsymbol{\Psi})\big)$. By use of Theorem 1.2, i.e., $\partial d_i(\boldsymbol{\vartheta}, \boldsymbol{\Psi})/\partial\boldsymbol{\vartheta} = \mathbb{E}(\boldsymbol{y}_i)$, we obtain the score vector as $\boldsymbol{u}(\boldsymbol{\vartheta}) = \frac{1}{n}\sum_{i=1}^{n}\big(\boldsymbol{y}_i - \mathbb{E}(\boldsymbol{y}_i)\big)$. The ML equations for the parameter $\boldsymbol{\vartheta}$ are therefore given by

$$\boldsymbol{u}(\hat{\boldsymbol{\vartheta}}) = \frac{1}{n}\sum_{i=1}^{n}\Big(\boldsymbol{y}_i - \mathbb{E}(\boldsymbol{y}_i)\Big) = \mathbf{0}\,,$$

and they are thus identical to

$$\frac{1}{n}\sum_{i=1}^{n}\boldsymbol{y}_i = \mathbb{E}\Big(\frac{1}{n}\sum_{i=1}^{n}\boldsymbol{y}_i\Big)\,. \tag{4.11}$$

Furthermore, as shown in the proof of Theorem 4.2, $\boldsymbol{B}(\boldsymbol{\beta}_0) = \mathbb{V}\mathrm{ar}(\boldsymbol{u}_i(\boldsymbol{\beta}_0))$.

Using the assumption of independence and Theorem 1.2, the matrix of second derivatives is given by

$$\boldsymbol{W}(\boldsymbol{\vartheta}) = -\frac{1}{n}\sum_{i=1}^{n}\frac{\partial^2 d_i(\boldsymbol{\vartheta})}{\partial\boldsymbol{\vartheta}\partial\boldsymbol{\vartheta}'} = -\frac{1}{n}\sum_{i=1}^{n}\mathbb{V}\mathrm{ar}(\boldsymbol{y}_i) = -n\mathbb{V}\mathrm{ar}\Big(\frac{1}{n}\sum_{i=1}^{n}\boldsymbol{y}_i\Big)\,.$$

For linear exponential families, the Hessian matrix of the natural parameter is identical to the negative Fisher information matrix. However, this is generally not true for the original parameter of distributions belonging to the linear exponential family, and this fact is illustrated in the next example.

Example 4.10 (Poisson distribution). Consider the Poisson distribution from Example 1.7. For n independently and identically $Po(\lambda)$ distributed random variables y_i, we have $\mathbb{E}(y_i) = \lambda = e^{\vartheta}$. The ML equations are therefore given by $\frac{1}{n}\sum_{i=1}^{n} y_i = e^{\hat{\vartheta}}$. Hence, $\ln\bar{y}$ is the ML estimator of ϑ, and, using the invariance principle of Theorem 4.4, $\hat{\lambda} = \frac{1}{n}\sum_{i=1}^{n} y_i = \bar{y}$ is the ML estimator of λ.

$\boldsymbol{A}_i(\vartheta) = -\mathbb{V}\mathrm{ar}(y_i) = -e^{\vartheta}$ and $\boldsymbol{B}_i(\vartheta) = \mathbb{E}(u_i(\vartheta)^2) = \mathbb{E}\big((y_i - \mathbb{E}(y_i))^2\big) = e^{\vartheta}$ are scalars. Analogously, $\boldsymbol{A}_i(\lambda)$ and $\boldsymbol{B}_i(\lambda)$ are scalars, and they can be obtained as follows. First- and second-order derivatives of the loglikelihood with respect to λ are given by

$$u_i(\lambda) = \frac{\partial l_i(\lambda)}{\partial\lambda} = \frac{y_i}{\lambda} - 1 \quad\text{and}\quad W_i(\lambda) = \frac{\partial^2 l_i(\lambda)}{\partial\lambda^2} = -\frac{y_i}{\lambda^2}\,, \tag{4.12}$$

respectively, yielding $\mathbb{E}(u_i) = 0$. Finally, the Fisher information is $\boldsymbol{A}_i(\lambda) = -\mathbb{E}\big(W_i(\lambda)\big) = \frac{1}{\lambda}$. The negative Hessian matrix using the parameterization in λ thus is different from the Fisher information.

Example 4.11 (Binomial distribution). $\vartheta = \mathrm{logit}(\pi)$ is the natural parameter of the binomial distribution (Example 1.8), and $\mathbb{E}(y) = n\pi$ if $y \sim B(n,\pi)$. The ML equations are therefore given by $y = n\hat{\pi}$. Hence, $\hat{\pi} = y/n$ is the ML estimator of π, and $\hat{\vartheta} = \ln(y/(n-y)$ is the ML estimator of ϑ.

Example 4.12 (Gamma distribution). Consider n independently identically $G(\alpha, \Psi)$ distributed random variables y_i with $\alpha > 0$ and fixed $\Psi > 0$ (Example 1.11). The natural parameter is $\vartheta = -\alpha$, and $\mathbb{E}(y_i) = -\frac{\Psi}{\vartheta}$. The ML equations for ϑ are therefore given by $\frac{1}{n}\sum_{i=1}^n y_i = -\Psi/\hat{\vartheta}$. Subsequently, the ML estimators of ϑ and α are given by $\hat{\vartheta} = -\Psi/\bar{y}$ and $\hat{\alpha} = \Psi/\bar{y}$, respectively.

Example 4.13 (Mean parameter of the multivariate normal distribution). Consider n independently identically multivariate normally distributed random variables $\boldsymbol{y}_i$, $i = 1, \ldots, n$, with mean vector $\boldsymbol{\mu}$ (Example 1.12). The ML equations for $\boldsymbol{\mu}$ are given by $\frac{1}{n}\sum_{i=1}^n \boldsymbol{y}_i = \hat{\boldsymbol{\mu}}$, and $\bar{\boldsymbol{y}}$ is the ML estimator of $\boldsymbol{\mu}$.

In the final example of this section, we derive the ML estimator for the variance of the univariate normal distribution. Here, the parameter of interest is not a function of the natural parameter from the simple linear exponential family. We therefore take the first derivative of the loglikelihood function to derive the ML estimator.

Example 4.14 (Variance of the univariate normal distribution). Consider n independently identically distributed random variables $y_i \sim N(\mu, \sigma^2)$ with $\sigma^2 > 0$. Then, the individual loglikelihood and the first derivative of the individual loglikelihood with respect to σ^2 are given by (Example 1.10)

$$l_i(\mu, \sigma^2) = -\tfrac{1}{2}\ln(2\pi) - \tfrac{1}{2}\ln(\sigma^2) - \frac{1}{2\sigma^2}(y_i - \mu)^2\,,$$

$$\frac{\partial l_i}{\partial \sigma^2} = -\frac{1}{2\sigma^2} + \frac{1}{2\sigma^4}(y_i - \mu)^2\,. \tag{4.13}$$

Subsequently, the ML equations are given by $\frac{1}{\hat{\sigma}^2} = \frac{1}{\hat{\sigma}^4}\frac{1}{n}\sum_{i=1}^n (y_i - \mu)^2$, and the ML estimator for σ^2 is

$$\hat{\sigma}^2 = \frac{1}{n}\sum_{i=1}^n (y_i - \mu)^2\,.$$

In almost all applications, the mean parameter μ is unknown, and it is therefore replaced by its estimator $\bar{y}$.

4.5 Maximum likelihood estimation in generalized linear models

In this section, we derive ML estimators for univariate and multivariate GLM.

4.5.1 Maximum likelihood estimation in univariate generalized linear models

As before, $\boldsymbol{\beta} = (\beta_1, \ldots, \beta_p)'$ is the p dimensional parameter vector of interest, the dependent variables $y_1, \ldots, y_n$ are collected in a vector $\boldsymbol{y}$, and, similarly, $\boldsymbol{X} = (\boldsymbol{x}_1, \ldots, \boldsymbol{x}_n)'$. Because observations are independent, the conditional covariance matrix of $\boldsymbol{y}$ given $\boldsymbol{X}$ is diagonal, i.e., $\boldsymbol{\Sigma} = \mathbb{V}\mathrm{ar}(\boldsymbol{y}|\boldsymbol{X}) = \mathrm{diag}\big(\mathbb{V}\mathrm{ar}(y_i|\boldsymbol{x}_i)\big)$. Furthermore, the $n \times n$ Jacobian of the link function $\boldsymbol{g}(\boldsymbol{\mu})$ is a diagonal matrix, i.e., $\partial \boldsymbol{g}'/\partial\boldsymbol{\mu} = \partial \boldsymbol{g}(\boldsymbol{\mu})'/\partial\boldsymbol{\mu} = \mathrm{diag}\big(\partial g(\mu_i)/\partial\mu_i\big)$. Note that $\boldsymbol{\mu} = \boldsymbol{\mu}(\boldsymbol{\beta})$ is a function of $\boldsymbol{\beta}$, and $\boldsymbol{g}(\boldsymbol{\mu}) = \boldsymbol{\eta}$.

Theorem 4.15. *Using the notations from above, the score vector of a univariate GLM is given by*

$$\boldsymbol{u}(\boldsymbol{\beta}) = \frac{\partial l(\boldsymbol{\beta})}{\partial \boldsymbol{\beta}} = \frac{1}{n}\boldsymbol{D}'\boldsymbol{\Sigma}^{-1}(\boldsymbol{y} - \boldsymbol{\mu})\,, \tag{4.14}$$

where $\boldsymbol{D} = \dfrac{\partial\boldsymbol{\mu}}{\partial\boldsymbol{\beta}'} = \dfrac{\partial\boldsymbol{\mu}}{\partial\boldsymbol{\eta}'}\dfrac{\partial\boldsymbol{\eta}}{\partial\boldsymbol{\beta}'} = \dfrac{\partial\boldsymbol{\mu}}{\partial\boldsymbol{g}(\boldsymbol{\mu})'}\boldsymbol{X}$ *is the matrix of first derivatives of* $\boldsymbol{\mu}$ *with respect to* $\boldsymbol{\beta}$. *The Hessian matrix and the Fisher information matrix are given by*

$$\boldsymbol{W}(\boldsymbol{\beta}) = \frac{\partial\boldsymbol{u}(\boldsymbol{\beta})}{\partial\boldsymbol{\beta}'} = -\frac{1}{n}\boldsymbol{X}'\left(\frac{\partial\boldsymbol{\mu}}{\partial\boldsymbol{g}(\boldsymbol{\mu})'}\boldsymbol{\Sigma}^{-1}\frac{\partial\boldsymbol{\mu}}{\partial\boldsymbol{g}(\boldsymbol{\mu})'} + diag\left((y_i - \mu_i)\frac{\partial^2\vartheta(\eta_i)}{\partial\eta^2}\right)\right)\boldsymbol{X}\,,$$

$$\text{and} \quad \boldsymbol{A}(\boldsymbol{\beta}) = \mathbb{E}^{\boldsymbol{X}}\,\mathbb{E}^{\boldsymbol{y}}(-\boldsymbol{W}(\boldsymbol{\beta})) = \mathbb{E}^{\boldsymbol{X}}\left(\frac{1}{n}\boldsymbol{D}'\boldsymbol{\Sigma}^{-1}\boldsymbol{D}\right), \tag{4.15}$$

respectively. If the natural link function is used, the expressions simplify to

$$\boldsymbol{u}(\boldsymbol{\beta}) = \frac{1}{n}\boldsymbol{X}'(\boldsymbol{y}-\boldsymbol{\mu}) \quad \text{and} \quad \boldsymbol{A}(\boldsymbol{\beta}) = \mathbb{E}^{\boldsymbol{X}}\big(-\boldsymbol{W}(\boldsymbol{\beta})\big) = \mathbb{E}^{\boldsymbol{X}}\left(\frac{1}{n}\boldsymbol{X}'\boldsymbol{\Sigma}^{-1}\boldsymbol{X}\right). \tag{4.16}$$

The functional relation between $\boldsymbol{\mu}$ and $\boldsymbol{\beta}$ is nonlinear, in general, so that $\hat{\boldsymbol{\beta}}$ cannot be calculated directly but has to be computed iteratively. Algorithms for this purpose, including the well-known Newton-Raphson and Fisher scoring algorithms, are described in detail, e.g., by Antoniou and Lu (2007).

Proof. Without loss of generality, we ignore the nuisance parameter $\boldsymbol{\Psi}$ in the proof. Then, $l(\vartheta_i) = \vartheta_i y_i - d(\vartheta_i)$ is the kernel of the individual loglikelihood function. With

$$\frac{\partial l_i(\boldsymbol{\beta})}{\partial\vartheta_i} = y_i - \frac{\partial d(\vartheta_i)}{\partial\vartheta} = y_i - \mu_i\,, \quad \frac{\partial\vartheta_i}{\partial\mu_i} = \mathbb{V}\mathrm{ar}(y_i|\boldsymbol{x}_i)\,, \quad \text{and} \quad \frac{\partial\mu_i}{\partial\eta_i} = \left(\frac{\partial g(\mu_i)}{\partial\mu}\right)^{-1},$$

we obtain

$$\frac{\partial l_i(\boldsymbol{\beta})}{\partial \beta_j} = \frac{\partial l_i(\boldsymbol{\beta})}{\partial \vartheta_i} \frac{\partial \vartheta_i}{\partial \mu_i} \frac{\partial \mu_i}{\partial \eta_i} x_{ij} = (y_i - \mu_i) \Big(\mathbb{V}\text{ar}(y_i|\boldsymbol{x}_i)\Big)^{-1} \left(\frac{\partial g(\mu_i)}{\partial \mu}\right)^{-1} x_{ij} \quad (4.17)$$

by use of the chain rule for $j = 1, \dots, p$. The matrix representation directly follows by summation. The matrix of second derivatives and the Fisher information matrix can be obtained similarly by noting that $\mathbb{E}^{y_i}(y_i - \mu_i) = 0$. The second term of the Hessian matrix thus equals $\mathbf{0}$ when the expected value is taken over $\boldsymbol{y}$. For natural link functions, $\theta_i = \eta_i$, thus $\partial\vartheta_i/\partial\eta_i = 1$ and $\partial^2\vartheta_i/\partial\eta_i^2 = 0$. □

4.5.2 Maximum likelihood estimation in multivariate generalized linear models

The dependent variables $\boldsymbol{y}_1, \dots, \boldsymbol{y}_n$ are collected in an $nT \times 1$ vector $\boldsymbol{y}$, and $\boldsymbol{X}_i$ are stacked to a matrix $\boldsymbol{X} = (\boldsymbol{X}_1', \boldsymbol{X}_2', \dots, \boldsymbol{X}_n')'$ of full rank. By writing $\boldsymbol{g}(\boldsymbol{\mu}) = (\boldsymbol{g}(\boldsymbol{\mu}_1)', \dots, \boldsymbol{g}(\boldsymbol{\mu}_n)')'$, we obtain $\boldsymbol{g}(\boldsymbol{\mu}) = \boldsymbol{X\beta}$. Finally, a block diagonal $nT \times nT$ covariance matrix $\boldsymbol{\Sigma}$ is defined comprising the covariance matrices $\boldsymbol{\Sigma}_i$ of the n independent clusters. Note that the covariance matrices $\boldsymbol{\Sigma}_i$ and the Jacobians $\partial\boldsymbol{g}'/\partial\boldsymbol{\mu}_i$ are generally not diagonal in the multivariate case.

Theorem 4.16. *With the notations from above, the score equations of a multivariate GLM are given by*

$$\boldsymbol{u}(\boldsymbol{\beta}) = \frac{\partial l(\boldsymbol{\beta})}{\partial \boldsymbol{\beta}} = \frac{1}{n}\boldsymbol{D}'\boldsymbol{\Sigma}^{-1}(\boldsymbol{y} - \boldsymbol{\mu}),$$

where $\boldsymbol{D} = \partial\boldsymbol{\mu}/\partial\boldsymbol{\beta}' = \boldsymbol{X}(\partial\boldsymbol{\mu}'/\partial\boldsymbol{\eta})$ *is the matrix of first derivatives. The Fisher information matrix is given by*

$$\boldsymbol{A}(\boldsymbol{\beta}) = \mathbb{E}^{\boldsymbol{X}}[\mathbb{E}^{\boldsymbol{y}}(-\boldsymbol{W}(\boldsymbol{\beta})) = \mathbb{E}^{\boldsymbol{X}}\left(\frac{1}{n}\boldsymbol{D}'\boldsymbol{\Sigma}^{-1}\boldsymbol{D}\right).$$

In case of natural link functions, expressions simplify to

$$\boldsymbol{u}(\boldsymbol{\beta}) = \frac{1}{n}\boldsymbol{X}'(\boldsymbol{y} - \boldsymbol{\mu}) \quad \text{and} \quad \boldsymbol{A}(\boldsymbol{\beta}) = \mathbb{E}^{\boldsymbol{X}}\left(\frac{1}{n}\boldsymbol{X}'\boldsymbol{\Sigma}^{-1}\boldsymbol{X}\right). \quad (4.18)$$

Proof. See, e.g., Fahrmeir and Tutz (2001, p. 105). □

4.6 Maximum likelihood estimation under misspecified models

In Sect. 4.2, we have seen that the correct specification of the model is the crucial assumption for Properties 3. and 4. of Theorem 4.2 to hold. Specifically, we have shown that the Fisher information matrix $\boldsymbol{A}$ needs to be equal to the OPG $\boldsymbol{B}$. This assumption is not necessarily fulfilled, as will be shown in the following example.

4.6.1 An example for model misspecification

In this section, we show that $-\boldsymbol{A}(\boldsymbol{\beta})$ need not equal $\boldsymbol{B}(\boldsymbol{\beta})$. We consider the estimation of both the mean and the variance of n independently identically random variables distributed as $y_i \sim N(\mu, \sigma^2)$ with $\sigma^2 > 0$. This corresponds to a linear regression model for y_i with $\boldsymbol{x}_i = 1$ for all i. The individual loglikelihood and its derivatives therefore are (see Examples 1.10 and 4.14)

$$l_i(\mu, \sigma^2) = l_i\big(\boldsymbol{\beta} = (\mu, \sigma^2)\big) = -\tfrac{1}{2}\ln(2\pi) - \tfrac{1}{2}\ln(\sigma^2) - \frac{1}{2\sigma^2}(y_i - \mu)^2\,, \tag{4.19}$$

$$\frac{\partial l_i}{\partial \mu} = \frac{1}{\sigma^2}(y_i - \mu)\,, \qquad \frac{\partial l_i}{\partial \sigma^2} = -\frac{1}{2\sigma^2} + \frac{1}{2\sigma^4}(y_i - \mu)^2\,,$$

$$\frac{\partial^2 l_i}{\partial \mu^2} = -\frac{1}{\sigma^2}\,, \quad \frac{\partial^2 l_i}{\partial (\sigma^2)^2} = \frac{1}{2\sigma^4} - \frac{1}{\sigma^6}(y_i - \mu)^2\,, \quad \frac{\partial^2 l_i}{\partial \mu \partial \sigma^2} = -\frac{1}{\sigma^4}(y_i - \mu)\,,$$

so that the Fisher information matrix and the OPG are given by

$$\boldsymbol{A}_i(\mu, \sigma^2) = \begin{pmatrix} 1/\sigma^2 & 0 \\ 0 & 1/2\sigma^4 \end{pmatrix}, \quad \text{and} \quad \boldsymbol{B}_i(\mu, \sigma^2) = \begin{pmatrix} 1/\sigma^2 & \gamma/2\sigma^3 \\ \gamma/2\sigma^3 & (\delta + 2)/4\sigma^4 \end{pmatrix},$$

respectively. Here, $\gamma = \mathbb{E}\big((x - \mu)^3\big)/\sigma^3$ denotes the skewness coefficient, and $\delta = \big(\mathbb{E}\big((x - \mu)^4\big)/\sigma^4\big) - 3$ is the degree of excess. The sandwich matrix $\boldsymbol{C}_i = \boldsymbol{A}_i^{-1}\boldsymbol{B}_i\boldsymbol{A}_i^{-1}$ of subject i is given by

$$\boldsymbol{C}_i(\mu, \sigma^2) = \begin{pmatrix} \sigma^2 & \gamma\,\sigma^3 \\ \gamma\,\sigma^3 & (\delta + 2)\sigma^4 \end{pmatrix}.$$

A necessary and sufficient condition for $\boldsymbol{A} = \boldsymbol{B}$ is $\gamma = 0$ and $\delta = 0$. This condition is fulfilled if the random variables are normally distributed. However, presence of skewness and/or kurtosis may lead to serious errors in inference.

In this example, we have seen that the model may easily be misspecified. Two questions now arise. First, what are the consequences of model misspecification? More specifically, how are inferences affected by model misspecification? This will be discussed in the next section. And second, are we able to detect model misspecification, e.g., by investigating differences between the Fisher information matrix $\boldsymbol{A}$ and the OPG $\boldsymbol{B}$? This will lead to the information matrix test, which will be introduced in Sect. 4.6.3. We start by giving an answer to the first question on the consequences of model misspecification.

4.6.2 Quasi maximum likelihood estimation

The key assumption of the previous sections was the correct specification of the true conditional density $f^{\star}(\boldsymbol{y}_i|\boldsymbol{X}_i||\boldsymbol{\beta})$ of $\boldsymbol{y}_i$ given $\boldsymbol{X}_i$, with $\boldsymbol{\beta}$ being the unknown parameter of interest. We will now study the effect of model misspecification. The basic theory for model misspecification, when the model is partly misspecified, was developed by Huber (1967), and it has been extended by White (1981; 1982) in his seminal papers (also see White, 1994).

By model misspecification we mean that there is no vector $\boldsymbol{\beta}_0$ such that the assumed density f is identical to the true density $f^{\star}$. More precisely, there is no vector $\boldsymbol{\beta}_0$ such that $f(\boldsymbol{y}_i|\boldsymbol{X}_i) = f^{\star}(\boldsymbol{y}_i|\boldsymbol{X}_i||\boldsymbol{\beta}_0)$. Even more, it might even be impossible to parameterize the true conditional density $f^{\star}$ in a parameter vector $\boldsymbol{\beta}$. As a consequence, the conditional moments of $\boldsymbol{y}_i$ given $\boldsymbol{X}_i$ under $f^{\star}$ generally do not coincide with the conditional moments of $\boldsymbol{y}_i$ given $\boldsymbol{X}_i$ under f if the model is misspecified.

This definition of model misspecification is very general, and it includes the misspecification of the entire distribution. For example, the true distribution might be a Cauchy distribution, while the one used in the statistical model is a normal distribution. In various settings, more specific definitions of misspecification have been used. For example, in GLM, over- or underdispersion is often encountered. Here, the mean structure might be correctly specified but the variance may be misspecified in the way that $\mathbb{V}\mathrm{ar}_f(y_i|\boldsymbol{x}_i||\boldsymbol{\beta}_0) \neq \mathbb{V}\mathrm{ar}_{f^{\star}}(y_i|\boldsymbol{x}_i||\boldsymbol{\beta}_0)$. A standard biometrical example with overdispersion is the number of boys born to each family. It should follow a binomial distribution but each family seems to skew the sex ratio of their children in favor of either boys or girls. There are not enough families close to the population 51:49 boy-to-girl mean ratio, and this yields an estimated variance that is larger than the one expected by the binomial model.

An even more specific type of misspecification in GLM is the misspecification of the link function (Li and Duan, 1989). Here, the true link function might be g^{-1}, while a different link function ${g^{\star}}^{-1}$ is used in the model. Other types of misspecification are related to the independent variables (Hall, 1999). For example, the distribution and the link function might have been correctly specified but some relevant independent variables have been omitted (Schmoor and Schumacher, 1997; Zhu et al., 2009) or the functional form of the independent variable might be wrong (Lin et al., 2002).

The key assumption of ML estimation is the correct specification of the true conditional density $f^\star(\boldsymbol{y}_i|\boldsymbol{X}_i)||\boldsymbol{\beta}_0)$ in the parameter of interest $\boldsymbol{\beta}_0$. This assumption includes that the link function, the functional form of the independent variables and the independent variables, are all correctly specified. If the distribution $f^\star$ is misspecified, we can still apply the standard ML approach. Under mild regularity conditions, there exists an ML estimator $\hat{\boldsymbol{\beta}}^\star$ even under the misspecified model, which is termed a quasi ML (QML) estimator because quasi is an ML estimator. More specifically, the sequence of maximized normed kernels of loglikelihood functions $l(\hat{\boldsymbol{\beta}})$ converges to the maximum of

$$\mathrm{E}_g^{\boldsymbol{X}}\Big(\mathrm{E}_{f^\star}^{\boldsymbol{y}} \ln f(\boldsymbol{y}|\boldsymbol{X}||\boldsymbol{\beta})\Big) = \int_{\mathcal{X}}\int_{\mathcal{Y}} \ln f(\boldsymbol{y}|\boldsymbol{X}||\boldsymbol{\beta}) f^\star(\boldsymbol{y}|\boldsymbol{X}) \boldsymbol{dy}\, g(\boldsymbol{X}) \boldsymbol{dX} \tag{4.20}$$

for all $\boldsymbol{\beta}$. This maximum is a function of $\boldsymbol{\beta}$, and the integration is taken over the true conditional density $f^\star$ rather than over the distribution f assumed by the researcher. Again, we stress that the true distribution $f^\star$ need not be parameterized in $\boldsymbol{\beta}$.

To interpret this maximum $\boldsymbol{\beta}^\star$, we utilize the Kullback-Leibler information criterion (KLIC; see, e.g., Arminger, 1995; Kullback and Leibler, 1951; White, 1994). It is given by

$$I\Big(f^\star(\boldsymbol{y}|\boldsymbol{X}), f(\boldsymbol{y}|\boldsymbol{X}||\boldsymbol{\beta})\Big) = \mathrm{E}_g^{\boldsymbol{X}}\Big(\mathrm{E}_{f^\star}^{\boldsymbol{y}} \ln f^\star(\boldsymbol{y}|\boldsymbol{X})\Big) - \mathrm{E}_g^{\boldsymbol{X}}\Big(\mathrm{E}_{f^\star}^{\boldsymbol{y}} \ln f(\boldsymbol{y}|\boldsymbol{X}||\boldsymbol{\beta})\Big) \tag{4.21}$$

and measures the average discrepancy between the true density $f^\star$ and the assumed density f for the parameter $\boldsymbol{\beta}$. With the use of Jensen's inequality (see, e.g., Rao, 1973, p. 59), it can be shown that $I(f, f^\star) \geq 0$ for any two densities f and $f^\star$. The KLIC equals 0 if and only if $f = f^\star$ almost surely. The first term on the right side of the KLIC from Eq. 4.21 is constant so that the KLIC is minimized if the second term is maximized. The QML estimator $\hat{\boldsymbol{\beta}}^\star$, i.e., the global unique maximum of the ML approach based on the density f, converges to $\boldsymbol{\beta}^\star$, the parameter value that minimizes the KLIC.

The QML estimator $\hat{\boldsymbol{\beta}}^\star$ is therefore still interpretable, even if the assumed density f is misspecified: It minimizes the discrepancy of the assumed density f and the true density $f^\star$. Because it minimally ignores the true structure, it has been called a "minimum ignorance estimator" (White, 1982, p. 4). In other words, in the sense of the KLIC, $\boldsymbol{\beta}^\star$ is the best possible approximation of the assumed density $f^\star$ to the true density f.

If the probability model is correctly specified, i.e., if the assumed density f and the true density $f^\star$ are equal for some value $\boldsymbol{\beta}_0$, then the KLIC $I(f, f^\star)$ attains its unique minimum at $\boldsymbol{\beta}^\star = \boldsymbol{\beta}_0$, so that $\hat{\boldsymbol{\beta}}$ is a consistent estimator for the true parameter vector $\boldsymbol{\beta}_0$. In this case, QML estimation is identical to ML estimation. Even if the assumed distribution f is misspecified, the asymptotic properties ML estimation of Theorem 4.2 can be adapted to QML estimation. For example, the QML estimator is asymptotically normally distributed with mean $\boldsymbol{\beta}^\star$. However, the Fisher information matrix $\boldsymbol{A}$ is no longer identical to the OPG $\boldsymbol{B}$. Specifically,

$$\sqrt{n}(\hat{\boldsymbol{\beta}} - \boldsymbol{\beta}^\star) \overset{a}{\sim} N\big(\mathbf{0}, [\boldsymbol{A}(\boldsymbol{\beta}^\star)]^{-1}\boldsymbol{B}(\boldsymbol{\beta}^\star)[\boldsymbol{A}(\boldsymbol{\beta}^\star)]^{-1}\big)\,. \tag{4.22}$$

An important consequence of White's (1982) findings is that estimation techniques are required that require less restricted assumptions about the model misspecification. In the next chapter, we will therefore consider an estimation approach where only the first moments, i.e., the mean structure are assumed to be correctly specified. Furthermore, in Chapts. 7 and 8, we will discuss models for the mean and the association structure, i.e., the first two moments, which require a correct specification of the first two moments only.

4.6.3 The information matrix test

In Sect. 4.6.1, we have seen that the crucial assumption of ML estimation is the equality of the Fisher information matrix $\boldsymbol{A}$ and the OPG $\boldsymbol{B}$. One approach for detecting model misspecification is therefore based on measuring the magnitude of the difference between $\boldsymbol{A}$ and $\boldsymbol{B}$. This difference may then serve as a basis for a formal statistical test. Because this test is based on the information matrix, it is termed the information matrix (IM) test for misspecification. The basic idea of the IM test was given by White (1982), and his work has been extended in several ways (see, e.g., Dhaene and Hoorelbeke, 2003; Hall, 1987; Horowitz, 1994; Lancaster, 1984; Orme, 1990; Stomberg and White, 2000; Zhang, 2001).

We start by considering the parametric model, where $l(\boldsymbol{\beta})$ denotes the joint loglikelihood function, and $\boldsymbol{\beta}$ is the $p \times 1$ parameter vector of interest. Furthermore, $\boldsymbol{\beta}_0$ maximizes $\mathbb{E}^{\boldsymbol{X}}\mathbb{E}^{\boldsymbol{y}}\big(l(\boldsymbol{\beta})\big)$ with respect to $\boldsymbol{\beta}$. Finally, $[\boldsymbol{u}_i]_l = [\boldsymbol{u}_i(\boldsymbol{\beta})]_l$ and $[\boldsymbol{u}_i]_m$ denote the lth and mth components of the score vector of individual i, respectively, and $[\boldsymbol{W}_i]_{lm} = [\boldsymbol{W}_i(\boldsymbol{\beta})]_{lm}$ is the lmth element of the Hessian matrix of subject i. The null hypothesis underlying all IM tests is H_0: $\boldsymbol{A} - \boldsymbol{B} = \mathbf{0}$, which is identical to

$$H_0\colon\ \mathbb{E}^{\boldsymbol{X}}\mathbb{E}^{\boldsymbol{y}}\Big([\boldsymbol{u}]_l[\boldsymbol{u}]_m + [\boldsymbol{W}]_{lm}\Big) = 0 \qquad \text{for} \qquad l, m = 1, \ldots, p\,.$$

Given a sample of observations $\boldsymbol{y}_i, \boldsymbol{X}_i$, we define the indicators

$$s_{i,lm} = [\boldsymbol{u}_i]_l[\boldsymbol{u}_i]_m + [\boldsymbol{W}_i]_{lm}\,, \qquad \text{and} \qquad s_{lm} = \tfrac{1}{n}\sum\nolimits_{i=1}^{n} s_{i,lm}\,,$$

which are based on the elements of the Hessian matrix and the outer product of the score vector.

IM tests rely on the idea that the estimated indicators $\hat{s}_{lm}$ will be jointly asymptotically normally distributed with mean 0 under the null hypothesis of no model misspecification under the regularity conditions for ML estimation. After derivation of the IM test statistic, it will be clear that some indicators automatically equal 0 in many applications, while others may be linear combinations of others (White, 1982). If we appropriately select a $q \times 1$ vector $\boldsymbol{s} = \boldsymbol{s}(\boldsymbol{\beta})$ of linearly independent indicators s_{lm}, we can construct a quadratic form $n\,\hat{\boldsymbol{s}}'\hat{\boldsymbol{V}}^{-1}\hat{\boldsymbol{s}}$, with $\hat{\boldsymbol{V}}$ being a regular estimate of the covariance matrix of $\boldsymbol{s}$ under H_0. The quadratic form then has an asymptotic χ^2 distribution with q degrees of freedom (d.f.). This leads to the following theorem:

Theorem 4.17. *Under H_0, the statistic $\sqrt{n}\hat{\boldsymbol{s}}$ is asymptotically normally distributed with mean $\boldsymbol{0}$ and covariance matrix*

$$\boldsymbol{V}(\boldsymbol{\beta}_0) = \mathbb{E}^{\boldsymbol{X}}\mathbb{E}^{\boldsymbol{y}}\Big(\big(\boldsymbol{s}(\boldsymbol{\beta}_0) - \boldsymbol{E}(\boldsymbol{\beta}_0)\boldsymbol{A}(\boldsymbol{\beta}_0)^{-1}\boldsymbol{u}(\boldsymbol{\beta}_0)\big) \\ \big(\boldsymbol{s}(\boldsymbol{\beta}_0) - \boldsymbol{E}(\boldsymbol{\beta}_0)\boldsymbol{A}(\boldsymbol{\beta}_0)^{-1}\boldsymbol{u}(\boldsymbol{\beta}_0)\big)'\Big),$$

where

$$\boldsymbol{E}(\boldsymbol{\beta}_0) = \mathbb{E}^{\boldsymbol{X}}\mathbb{E}^{\boldsymbol{y}}\Big(\frac{\partial \boldsymbol{s}(\boldsymbol{\beta}_0)}{\partial \boldsymbol{\beta}'}\Big).$$

The IM test can be carried out using the Wald statistic

$$\widehat{IM} = n\,\hat{\mathbf{s}}(\hat{\boldsymbol{\beta}})'\hat{\mathbf{V}}(\hat{\boldsymbol{\beta}})^{-1}\hat{\mathbf{s}}(\hat{\boldsymbol{\beta}}),$$

where $\hat{\mathbf{V}}(\hat{\boldsymbol{\beta}})$ is a consistent estimator of $\mathbf{V}(\boldsymbol{\beta}_0)$. $\widehat{IM}$ is asymptotically centrally χ^2 distributed with q d.f. under H_0. The hypothesis of a correctly specified model is rejected for large values of $\widehat{IM}$. $\mathbf{V}(\boldsymbol{\beta}_0)$ can be estimated consistently by explicitly calculating the expected value of $\mathbf{V}(\boldsymbol{\beta}_0)$ and then replacing $\boldsymbol{\beta}$ with its estimator. Alternatively, $\mathbf{V}(\boldsymbol{\beta}_0)$ may be estimated by

$$\hat{\boldsymbol{V}}(\hat{\boldsymbol{\beta}}) = \frac{1}{n}\sum_{i=1}^{n}\Big(\big(\hat{\boldsymbol{s}}(\hat{\boldsymbol{\beta}}) - \hat{\boldsymbol{E}}(\hat{\boldsymbol{\beta}})\hat{\boldsymbol{A}}(\hat{\boldsymbol{\beta}})^{-1}\hat{\boldsymbol{u}}(\hat{\boldsymbol{\beta}})\big)\big(\hat{\boldsymbol{s}}(\hat{\boldsymbol{\beta}}) - \hat{\boldsymbol{E}}(\hat{\boldsymbol{\beta}})\hat{\boldsymbol{A}}(\hat{\boldsymbol{\beta}})^{-1}\hat{\boldsymbol{u}}(\hat{\boldsymbol{\beta}})\big)'\Big). \tag{4.23}$$

Remark 4.18. Because the convergence to the asymptotic distribution is rather slow, the use of jack-knife or bootstrap procedures has been proposed for estimating the covariance matrix $\boldsymbol{V}(\boldsymbol{\beta}_0)$ (see, e.g., Stomberg and White, 2000; Dhaene and Hoorelbeke, 2003).

Proof. The first proof has been given by White (1982, Theorem 4.1), and detailed proofs can be found, e.g., in White (1994) or in Gourieroux and Monfort (1995b). The idea of all proofs is to use a first-order Taylor series of $\sqrt{n}\hat{\boldsymbol{s}}(\hat{\boldsymbol{\beta}})$ around $\boldsymbol{\beta}_0$ in a first step and to employ the central limit theorem in the second. □

The presented IM test is based on a first-order Taylor series. Therefore, both the approximation of the IM test to its finite sample distribution and its power may be improved by using a second-order IM test that relies on a second-order Taylor series.

At the end of this chapter, we give three examples for the IM test.

Example 4.19 (Information matrix test for the Poisson distribution). When considering data that might follow a Poisson distribution, we usually compare the empirical mean and the empirical variance because they should be very similar. If the variance is substantially larger or smaller than the mean, we call the model overdispersed and underdispersed, respectively. With the IM test, we are able to formally test the hypothesis of a misspecified Poisson model.

Consider a sample of n independently and identically $Po(\lambda)$ distributed random variables. The kernel of the individual loglikelihood function is given by $l_i(\lambda) = y_i \ln(\lambda) - \lambda$ (Example 4.10). The score vector and the Hessian matrix are scalars, and they are given by $u_i(\lambda) = \frac{y_i}{\lambda} - 1$ and $W_i(\lambda) = -\frac{y_i}{\lambda^2}$, respectively (Eq. 4.12), yielding

$$s_i(\lambda) = \frac{y_i^2}{\lambda^2} - \frac{y_i}{\lambda^2} - \frac{2y_i}{\lambda} + 1 \quad \text{and} \quad s(\lambda) = \frac{\overline{y^2}}{\lambda^2} - \frac{\bar{y}}{\lambda^2} - \frac{2\bar{y}}{\lambda} + 1\,. \tag{4.24}$$

If the ML estimator $\hat{\lambda} = \bar{y}$ of λ is used for estimating $s(\lambda)$, Eq. 4.24 reduces to

$$\hat{s}(\hat{\lambda}) = \frac{\overline{y^2}}{\bar{y}^2} - \frac{1}{\bar{y}} - 1 = \frac{\overline{y^2} - \bar{y}^2 - \bar{y}}{\bar{y}^2}\,.$$

Because $\overline{y^2} - \bar{y}^2$ is an estimator of the variance, and $\bar{y}$ is an estimator of the mean, $\hat{s}(\hat{\lambda})$ measures whether the variance equals the mean. With the denominator $\bar{y}^2$, $\hat{s}$ is independent of dimensions. It is used for standardization, analogously to the arithmetic mean in the coefficient of variation.

The IM test statistic $\widehat{IM}$ is easily computed from s and the expression for $E(s(\lambda_0))$. In fact, $E(s(\lambda_0)) = 0$ because $\frac{\partial s_i(\lambda)}{\partial \lambda} = 2\lambda^{-2}(-y_i^2\lambda^{-1} + y_i\lambda^{-1} + y_i)$, $\mathbb{E}(y_i) = \lambda$, and $\mathbb{E}(y_i^2) = \lambda^2 + \lambda$.

Subsequently, $\widehat{IM}$ reduces to $n\,\hat{s}^2 / \frac{1}{n}\sum_{i=1}^n \hat{s_i}^2 = n\big(\frac{1}{n}\sum_{i=1}^n \hat{s_i}\big)^2 / \big(\frac{1}{n}\sum_{i=1}^n \hat{s_i}^2\big)$ if $V(\lambda)$ is estimated by Eq. 4.23, and $\widehat{IM}$ is asymptotically χ^2 distributed with 1 d.f. under H_0.

A simpler alternative for the IM test statistic is $\widehat{IM} = n\,\hat{s}^2\,\bar{y}^2/2$, which can be obtained using $\mathbb{V}\text{ar}(\lambda_0) = \mathbb{E}(s_i^2) = 2/\lambda_0^2$ because of $\mathbb{E}((y_i - \lambda)^3) = \lambda$, and $\mathbb{E}((y_i - \lambda)^4) = 3\lambda^2 + \lambda$. This test statistic is also asymptotically χ^2 distributed with 1 d.f. under H_0.

Example 4.20 (Information matrix test for the mean model). Consider the simple linear regression model $y_i = \mu + \varepsilon_i$, where the errors ε_i are independently and identically normally distributed with variance σ^2. The kernel of the individual loglikelihood function as well as the individual score vector and

Hessian matrix have been given in Eq. 4.19. With $\boldsymbol{\xi} = (\mu, \sigma^2)'$, we obtain

$$\boldsymbol{s}_i(\boldsymbol{\xi}) = \frac{1}{\sigma^2} \begin{pmatrix} \frac{1}{\sigma^2}(y_i - \mu)^2 - 1 \\ -\frac{1}{2}(y_i - \mu) + \frac{1}{2\sigma^2}(y_i - \mu)^3 - \frac{1}{\sigma^2}(y_i - \mu) \\ \frac{1}{4\sigma^2} + \frac{1}{4\sigma^6}(y_i - \mu)^4 - \frac{1}{2\sigma^4}(y_i - \mu)^2 + \frac{1}{2\sigma^2} - \frac{1}{\sigma^4}(y_i - \mu)^2 \end{pmatrix}$$

and

$$\hat{\boldsymbol{s}}(\hat{\boldsymbol{\xi}}) = \frac{1}{\hat{\sigma}^2} \begin{pmatrix} 0 \\ \frac{1}{2\hat{\sigma}^2} \sum_{i=1}^n (y_i - \hat{\mu})^3 \\ -\frac{3}{4\hat{\sigma}^2} + \frac{1}{4\hat{\sigma}^6} \sum_{i=1}^n (y_i - \hat{\mu})^4 \end{pmatrix} .$$

Because the first component of $\hat{\boldsymbol{s}}(\hat{\boldsymbol{\beta}})$ equals 0, the IM test statistic does not measure differences between mean and variance. Instead, it tests whether the skewness $\mathbb{E}((y_i - \mu)^3)/\sigma^3$ and the coefficient of kurtosis $\big(\mathbb{E}((y_i - \mu)^4)/\sigma^4\big) - 3$ equal 0, which should be true for the normal distribution.

Example 4.21 (Information matrix test for the linear regression model). Consider the linear regression model $y_i|\boldsymbol{x}_i \sim N(\boldsymbol{x}_i'\boldsymbol{\beta}, \sigma^2)$ for all i, i.e., the observations are independent, and the conditional distribution of y_i given $\boldsymbol{x}_i$ follows a normal distribution with homoscedastic errors σ^2. Let the parameter vector of interest be $\boldsymbol{\xi} = (\boldsymbol{\beta}', \sigma^2)'$.

With the kernel of the individual loglikelihood function given by $l(\boldsymbol{\xi}) = -\frac{1}{2}\ln\sigma^2 - \frac{(y_i - \boldsymbol{x}_i'\boldsymbol{\beta})^2}{2\sigma^2}$, we obtain

$$\boldsymbol{u}_i(\boldsymbol{\xi}) = \begin{pmatrix} \frac{\boldsymbol{x}_i\varepsilon_i}{\sigma^2} \\ -\frac{1}{2\sigma^2} + \frac{\varepsilon_i^2}{2\sigma^4} \end{pmatrix} \quad \text{and} \quad \boldsymbol{W}_i(\boldsymbol{\xi}) = \begin{pmatrix} -\frac{\boldsymbol{x}_i\boldsymbol{x}_i'}{\sigma^2} & -\frac{\boldsymbol{x}_i\varepsilon_i}{\sigma^4} \\ -\frac{\varepsilon_i\boldsymbol{x}_i'}{\sigma^4} & \frac{1}{2\sigma^4} - \frac{\varepsilon_i^2}{\sigma^6} \end{pmatrix}$$

after some algebra. Here, $\varepsilon_i = y_i - \boldsymbol{x}_i'\boldsymbol{\beta}$ denotes the ordinary residual.

The QML estimator is obtained by setting the average of the normed score vector to $\boldsymbol{0}$ and solving for both $\boldsymbol{\beta}$ and σ^2. The QML estimator of $\boldsymbol{\beta}$ is the ordinary least squares estimator, i.e., $\hat{\boldsymbol{\beta}} = \big(\sum_{i=1}^n \boldsymbol{x}_i\boldsymbol{x}_i'\big)^{-1} \sum_{i=1}^n \boldsymbol{x}_i'y_i$, and the QML estimator of σ^2 is the average of the ordinary least squares residuals, i.e., $\hat{\sigma}^2 = \frac{1}{n}\sum_{i=1}^n (y_i - \boldsymbol{x}_i'\hat{\boldsymbol{\beta}})^2 = \frac{1}{n}\sum_{i=1}^n \hat{\varepsilon}_i^2$.

The outer product is given by

$$\boldsymbol{u}_i(\boldsymbol{\xi})\boldsymbol{u}_i(\boldsymbol{\xi})' = \begin{pmatrix} \frac{\varepsilon_i^2\boldsymbol{x}_i\boldsymbol{x}_i'}{\sigma^4} & -\frac{\boldsymbol{x}_i\varepsilon_i}{2\sigma^4} + \frac{\boldsymbol{x}_i\varepsilon_i^3}{2\sigma^6} \\ -\frac{\varepsilon_i\boldsymbol{x}_i'}{2\sigma^4} + \frac{\varepsilon_i^3\boldsymbol{x}_i'}{2\sigma^6} & \frac{1}{4\sigma^4} - \frac{\varepsilon_i^2}{2\sigma^6} + \frac{\varepsilon_i^4}{4\sigma^8} \end{pmatrix},$$

and the IM test statistic therefore consists of three relevant matrix blocks. While the top left block contains $p(p+1)/2$ different elements, yielding a $p(p+1)/2 \times 1$ vector of indicators, the top right and bottom right block consist of p and 1 elements, respectively.

A typical top left component of the vector of indicators is given by

$$\boldsymbol{s}_{i,lm}(\boldsymbol{\xi}) = [\boldsymbol{u}_{i1}]_l[\boldsymbol{u}_{i1}]_m + [\boldsymbol{W}_{i11}]_{lm} = x_{il}x_{im}\frac{\varepsilon_i^2 - \sigma^2}{\sigma^4},$$

resulting in $\hat{\boldsymbol{s}}_{lm}(\hat{\boldsymbol{\xi}}) = \frac{1}{n}\sum_{i=1}^n x_{il}x_{im}(\hat{\varepsilon}_i^2 - \hat{\sigma}^2)/\hat{\sigma}^4$. Similarly, one obtains $\boldsymbol{s}_{i,m}(\boldsymbol{\xi}) = \frac{x_{im}\varepsilon_i^3}{2\sigma^6} - \frac{3}{2}\frac{x_{im}\varepsilon_i}{\sigma^4}$ for the top right block, and this gives $\hat{s}_m(\hat{\boldsymbol{\xi}}) = \frac{1}{n}\sum_{i=1}^n x_{im}\hat{\varepsilon}_i^3/(2\hat{\sigma}^6)$ because $\boldsymbol{X}'\hat{\boldsymbol{\varepsilon}} = \boldsymbol{0}$. Finally, for the low right block we have $s_i(\boldsymbol{\xi}) = \frac{3}{4\sigma^4} - \frac{6\varepsilon_i^2}{4\sigma^4} + \frac{\varepsilon_4}{4\sigma^8}$, which results in $\hat{\boldsymbol{s}}(\hat{\boldsymbol{\xi}}) = \frac{1}{n}\sum_{i=1}^n(\hat{\varepsilon}_i^4 - 3\hat{\sigma}^4)/(4\hat{\sigma}^8)$.

Subsequent calculations show (see, e.g., Hall, 1987, 1989) that the IM test statistic for the linear model is asymptotically equivalent to the sum of the three test statistics

$$T_1 = \left(\sum_{i=1}^n(\hat{\varepsilon}_i^2 - \hat{\sigma}^2)\varsigma_i'\right)\mathbf{1}_q\left(\mathbf{1}_q'\sum_{i=1}^n(\varsigma_i\varsigma_i')\mathbf{1}_q\right)^{-1}\mathbf{1}_q'\left(\sum_{i=1}^n(\hat{\varepsilon}_i^2 - \hat{\sigma}^2)\varsigma_i\right)\Big/(2\hat{\sigma}^4),$$
$$T_2 = \left(\sum_{i=1}^n\hat{\varepsilon}_i^3\boldsymbol{x}_i'\right)\mathbf{1}_p\left(\mathbf{1}_p'\sum_{i=1}^n\boldsymbol{x}_i\boldsymbol{x}_i'\mathbf{1}_p'\right)^{-1}\mathbf{1}_p'\left(\sum_{i=1}^n\hat{\varepsilon}_i^3\boldsymbol{x}_i\right)/(6\hat{\varepsilon}_i^6),$$
$$T_3 = \frac{1}{n}\left(\sum_{i=1}^n(\hat{\varepsilon}_i^4 - 3\hat{\sigma}^4)\right)^2/(24\hat{\sigma}^8),$$

where $\mathbf{1}_k = (1,\ldots,1)'$ denotes the 1-vector of length k, $q = p(p+1)/2$, and ς_i is a $p(p+1)/2$ vector consisting of the lower triangular elements of $\boldsymbol{x}_i\boldsymbol{x}_i' - \frac{1}{n}\sum_{i=1}^n\boldsymbol{x}_i\boldsymbol{x}_i'$. Under H_0 of no model misspecification, the test statistic $\widehat{IM} = \sum_{j=1}^3 T_j$ is asymptotically χ^2 distributed with $(p^2(p+1))/(2+p+1)$ d.f.

The first test statistic T_1 is a test for heteroscedasticity because it measures differences between the residuals ε_i and the common variance σ^2 (see, e.g., Hall, 1987; White, 1980). The second and third statistics have flavors of test statistics for skewness and kurtosis, respectively, in the linear model with the assumption of normality as seen, e.g., in the previous example. The IM tests T_2 and T_3 are therefore asymptotically equivalent to a series of other tests if the matrix of explanatory variables only consists of a regression constant. These connections have been described in detail by Hall (1987, 1989).

It is interesting to see that T_1 to T_3 are sensitive to deviations from the normal distribution, but none of them is sensitive to serial correlation. If the aim is to detect serial correlation, Hausman-type tests of misspecification could be employed (Hausman, 1978; Holly, 1982).

Chapter 5
Pseudo maximum likelihood method based on the linear exponential family

In the previous chapter, we discussed the classical ML approach, its asymptotic properties, and the effect of model misspecification. In this chapter, we consider a generalization of the ML method that explicitly allows for partial model misspecification. Here, the idea of ML estimation is still employed, but the ML technique is applied to a probably misspecified density from the linear exponential family. Therefore, this estimation approach is termed pseudo maximum likelihood (PML) estimation. PML estimation was introduced by Gourieroux et al. in their seminal papers (1984a, 1984b), it has been reviewed, e.g., in Arminger (1995) and Gourieroux and Monfort (1993, 1995a), and it has been generalized in several ways; see, e.g., Broze and Gourieroux (1998) or Magnus (2007).

In this chapter, we consider PML estimation, where only the mean structure needs to be correctly specified, and it is subsequently termed PML1 estimation. We show that under mild regularity conditions, the parameter vector of the mean structure can be consistently estimated. Furthermore, we show that it is asymptotically normally distributed. Although PML1 estimation has the advantage of guaranteeing consistent parameter estimates even if the model is partly misspecified, it has the disadvantage of being less efficient than ML estimation when the model is correctly specified. The robust variance estimator is of the sandwich type, and it therefore includes the product of three terms with the Fisher information being the bread and the OPG being the butter. Because of the multiplication, the robust variance is biased, and it is less stable–Kauermann and Carroll (2001) call it extra variability–than the standard ML variance estimator, which is the inverse Fisher information. Therefore, small sample properties and adjustments for improvement need to be discussed in some detail.

The chapter is organized as follows. We first define the PML1 estimator, and second, derive its asymptotic properties (Sect. 5.2). We next illustrate the PML1 approach in a series of examples (Sect. 5.3). Specifically, we consider the linear regression model with heteroscedastic errors, the independence estimating equations (IEE) with the covariance matrix being equal to the identity matrix, and the generalized estimating equations (GEE1) with fixed covariance matrix. In Sect. 5.4, we finally compare the efficiency of the PML1 method with the ML approach, consider bias corrections of the robust variance estimator, and discuss small sample adjustments for the robust variance estimator.

5.1 Definition

Consider a sample of n independently distributed T-dimensional random vectors $\boldsymbol{y}_i$, and $\boldsymbol{X}_i$ is the $T \times p$ matrix of stochastic and/or fixed explanatory variables of subject i. The true but unknown density (or probability mass function for discrete random vectors) of $\boldsymbol{y}_i$ given $\boldsymbol{X}_i$ is denoted by $f^\star(\boldsymbol{y}_i|\boldsymbol{X}_i)$ with conditional expectation $\mathbb{E}_{f^\star}(\boldsymbol{y}_i|\boldsymbol{X}_i)$ and conditional covariance matrix $\mathbb{V}\mathrm{ar}_{f^\star}(\boldsymbol{y}_i|\boldsymbol{X}_i) = \boldsymbol{\Omega}^\star(\boldsymbol{X}_i)$.

The true density $f^\star$ may be different from the assumed or pseudo density f specified by the researcher. More specifically, the assumed conditional density f of $\boldsymbol{y}_i$ given $\boldsymbol{X}_i$ is parameterized in the $p \times 1$ parameter vector $\boldsymbol{\beta}$ so that the conditional mean of $\boldsymbol{y}_i$ given $\boldsymbol{X}_i$ depends on $\boldsymbol{\beta}$, i.e., $\mathbb{E}_f(\boldsymbol{y}_i|\boldsymbol{X}_i||\boldsymbol{\beta})$. If a vector $\boldsymbol{\beta}_0$ exists such that the mean of the true distribution equals the mean of the assumed distribution, i.e., if

$$\mathbb{E}_{f^\star}(\boldsymbol{y}_i|\boldsymbol{X}_i) = \mathbb{E}_f(\boldsymbol{y}_i|\boldsymbol{X}_i||\boldsymbol{\beta}_0)\,, \tag{5.1}$$

then $f^\star$ is partially parameterized in $\boldsymbol{\beta}_0$.

Throughout this chapter we assume that Eq. 5.1 holds. However, the true conditional variance matrix $\boldsymbol{\Omega}^\star(\boldsymbol{X}_i)$ of $\boldsymbol{y}_i$ given $\boldsymbol{X}_i$ need not be parameterized in $\boldsymbol{\beta}$. The stochastic model of $\boldsymbol{y}_i$ given $\boldsymbol{X}_i$ under the true model $f^\star(\boldsymbol{y}_i|\boldsymbol{X}_i||\boldsymbol{\beta}_0)$ is therefore given by

$$\begin{aligned}
\boldsymbol{y}_i &= \boldsymbol{\mu}(\boldsymbol{X}_i, \boldsymbol{\beta}_0) + \boldsymbol{\varepsilon}_i^\star \qquad \text{with} \qquad \mathbb{E}(\boldsymbol{\varepsilon}_i^\star|\boldsymbol{X}_i) = \boldsymbol{0}\,,\\
\mathbb{E}_{f^\star}(\boldsymbol{y}_i|\boldsymbol{X}_i) &= \mathbb{E}_f(\boldsymbol{y}_i|\boldsymbol{X}_i) = \boldsymbol{\mu}(\boldsymbol{X}_i, \boldsymbol{\beta}_0) = \boldsymbol{\mu}_i\,,\\
\mathbb{V}\mathrm{ar}_{f^\star}(\boldsymbol{y}_i|\boldsymbol{X}_i) &= \mathbb{V}\mathrm{ar}(\boldsymbol{\varepsilon}_i^\star|\boldsymbol{X}_i) = \boldsymbol{\Omega}(\boldsymbol{X}_i)\,.
\end{aligned}$$

Again, we stress that the true conditional density $f^\star$ possibly depends on the parameter vector $\boldsymbol{\beta}_0$ only through the mean structure $\boldsymbol{\mu}(\boldsymbol{X}_i, \boldsymbol{\beta}_0)$. In GEE, the mean structure is commonly chosen by using a link function from the generalized linear model. Specifically, if we use the notation of Definition 3.5, this choice gives $\boldsymbol{\mu}(\boldsymbol{X}_i, \boldsymbol{\beta}) = \boldsymbol{g}(\boldsymbol{\eta}_i) = \boldsymbol{g}(\boldsymbol{X}_i\boldsymbol{\beta})$. Furthermore, we assume the existence of the variance matrix $\boldsymbol{\Omega}(\boldsymbol{X}_i)$, but the correct specification of the covariance matrix is not required for PML1 estimation.

Furthermore, we assume that the pseudo distribution belongs to the linear exponential family with fixed nuisance parameter. The stochastic model of $\boldsymbol{y}_i$ given $\boldsymbol{X}_i$ under the assumed model $f(\boldsymbol{y}_i|\boldsymbol{X}_i||\boldsymbol{\beta}_0)$ is thus given by

$$\begin{aligned}
\boldsymbol{y}_i &= \boldsymbol{\mu}(\boldsymbol{X}_i, \boldsymbol{\beta}_0) + \boldsymbol{\varepsilon}_i \qquad \text{with} \qquad \mathbb{E}(\boldsymbol{\varepsilon}_i|\boldsymbol{X}_i) = \boldsymbol{0}\,, \\
\mathbb{E}_{f^\star}(\boldsymbol{y}_i|\boldsymbol{X}_i) &= \mathbb{E}_f(\boldsymbol{y}_i|\boldsymbol{X}_i) = \boldsymbol{\mu}(\boldsymbol{X}_i, \boldsymbol{\beta}_0) = \boldsymbol{\mu}_i\,, \\
\mathbb{V}\mathrm{ar}_f(\boldsymbol{y}_i|\boldsymbol{X}_i) &= \mathbb{V}\mathrm{ar}(\boldsymbol{\varepsilon}_i|\boldsymbol{X}_i) = \boldsymbol{\Sigma}(\boldsymbol{X}_i, \boldsymbol{\beta}, \boldsymbol{\Psi}_i)\,.
\end{aligned}$$

Because the assumed density belongs to the linear exponential family, the kernel of the normed loglikelihood function of the assumed distribution is given by

$$l(\boldsymbol{\beta}) = \frac{1}{n}\sum_{i=1}^{n}\Big(a\big(\boldsymbol{\mu}(\boldsymbol{X}_i,\boldsymbol{\beta}),\boldsymbol{\Psi}_i\big) + \boldsymbol{c}\big(\boldsymbol{\mu}(\boldsymbol{X}_i,\boldsymbol{\beta}),\boldsymbol{\Psi}_i\big)'\boldsymbol{y}_i\Big)\,. \tag{5.2}$$

Estimation is based on the assumed density, i.e., the pseudo density, and $l(\boldsymbol{\beta})$ is therefore usually termed normed pseudo loglikelihood function.

Additionally, we assume first-order identifiability of the parameter vector $\boldsymbol{\beta}$, i.e., if $\boldsymbol{\mu}(\boldsymbol{X}, \boldsymbol{\beta}_1) = \boldsymbol{\mu}(\boldsymbol{X}, \boldsymbol{\beta}_2)$ implies $\boldsymbol{\beta}_1 = \boldsymbol{\beta}_2$ almost surely. Another important assumption is related to the domain of the conditional expectations of the true density and the pseudo density. Specifically, the domain of the conditional expectation of the true density $\mathbb{E}_{f^\star}(\boldsymbol{y}_i|\boldsymbol{X}_i||\boldsymbol{\beta}_0)$ needs to be a subset of the domain of the conditional expectation $\mathbb{E}_f(\boldsymbol{y}_i|\boldsymbol{X}_i||\boldsymbol{\beta}_0)$ of the pseudo density.

The need for this assumption is best explained by examples: If the univariate dependent random variable y_i can take negative values so that $\mathbb{E}_{f^\star}(y_i|\boldsymbol{x}_i) < 0$ is possible, no pseudo distribution with the restriction $\mu > 0$ should be chosen. As a result, the normal distribution is generally used as pseudo distribution in this case. The normal distribution can also be used for any continuous or discrete dependent variables. In contrast, the Poisson distribution might be chosen as pseudo distribution if $y_i > 0$ throughout. The Poisson distribution can be used as pseudo distribution in this case even if y_i is a continuous random variable. To repeat, the important restriction in this example is that $\mathbb{E}_{f^\star}(\boldsymbol{y}_i|\boldsymbol{X}_i)$ is positive for any y_i. To give another example, the Poisson distribution might be used as pseudo distribution if the true distribution is binomial. However, the binomial distribution is not a valid choice if the true distribution is Poisson because the mean of the binomial distribution is restricted to the interval $[0; 1]$, but the intensity parameter of the Poisson distribution can take any positive real value.

Now we can define the PML1 estimator using the kernel of the normed pseudo loglikelihood from Eq. 5.2:

Definition 5.1 (PML1 estimator). A pseudo maximum likelihood estimator for the mean structure or, briefly, PML1 estimator of $\boldsymbol{\beta}$ is any value $\hat{\boldsymbol{\beta}}$ maximizing the kernel of the normed pseudo loglikelihood function

$$l(\boldsymbol{\beta}) = \frac{1}{n}\sum_{i=1}^{n}\Big(a\big(\boldsymbol{\mu}(\boldsymbol{X}_i,\boldsymbol{\beta}),\boldsymbol{\Psi}_i\big) + \boldsymbol{c}\big(\boldsymbol{\mu}(\boldsymbol{X}_i,\boldsymbol{\beta}),\boldsymbol{\Psi}_i\big)'\boldsymbol{y}_i\Big)\,.$$

5.2 Asymptotic properties

Let $\boldsymbol{D}_i = \partial \boldsymbol{\mu}_i / \partial \boldsymbol{\beta}'$ denote the matrix of first partial derivatives of the mean with respect to $\boldsymbol{\beta}$. Under the assumptions of the previous section and the standard ML regularity conditions (see, e.g., Gourieroux et al., 1984b; White, 1982), we can show the following asymptotic statements for PML1 estimation.

Theorem 5.2 (Properties of PML1 estimators).

1. *There asymptotically exists a PML1 estimator $\hat{\boldsymbol{\beta}}$ for the true parameter vector $\boldsymbol{\beta}_0$.*
2. *The PML1 estimator $\hat{\boldsymbol{\beta}}$ converges almost surely to the true parameter $\boldsymbol{\beta}_0$.*
3. *The score vector for $\boldsymbol{\beta}$ is given by*
$$\boldsymbol{u}(\boldsymbol{\beta}) = \sum_{i=1}^{n} \boldsymbol{D}_i' \boldsymbol{\Sigma}_i^{-1} (\boldsymbol{y}_i - \boldsymbol{\mu}_i) = \boldsymbol{D}' \boldsymbol{\Sigma}^{-1} (\boldsymbol{y} - \boldsymbol{\mu}) \,,$$
where $\boldsymbol{D}$ is the stacked matrix of the $\boldsymbol{D}_i$, $\boldsymbol{\Sigma}$ is the block diagonal matrix of the $\boldsymbol{\Sigma}_i$, and $\boldsymbol{y}$ and $\boldsymbol{\mu}$ are the stacked vectors $\boldsymbol{y}_i$ and $\boldsymbol{\mu}_i$, respectively.
4. *The PML1 estimator $\hat{\boldsymbol{\beta}}$ for $\boldsymbol{\beta}_0$ is asymptotically normal. More specifically,*
$$\sqrt{n}(\hat{\boldsymbol{\beta}} - \boldsymbol{\beta}_0) \overset{a}{\sim} N\Big(\boldsymbol{0}, \boldsymbol{A}(\boldsymbol{\beta}_0)^{-1} \boldsymbol{B}(\boldsymbol{\beta}_0) \boldsymbol{A}(\boldsymbol{\beta}_0)^{-1}\Big) = N\Big(\boldsymbol{0}, \boldsymbol{C}(\boldsymbol{\beta}_0)^{-1}\Big) \,, \quad (5.3)$$
where $\boldsymbol{A}(\boldsymbol{\beta}) = \mathbb{E}^{\boldsymbol{X}}\big(\mathbb{E}^{\boldsymbol{y}}_{f^\star} - \boldsymbol{W}_i(\boldsymbol{\beta})\big) = \mathbb{E}^{\boldsymbol{X}}(\boldsymbol{D}_i' \boldsymbol{\Sigma}_i^{-1} \boldsymbol{D}_i)$ is the Fisher information matrix and $\boldsymbol{B}(\boldsymbol{\beta}) = \mathbb{E}^{\boldsymbol{X}} \mathbb{E}^{\boldsymbol{y}} (\boldsymbol{u}_i(\boldsymbol{\beta}) \boldsymbol{u}_i(\boldsymbol{\beta})') = \mathbb{E}^{\boldsymbol{X}}(\boldsymbol{D}_i' \boldsymbol{\Sigma}_i^{-1} \boldsymbol{\Omega}_i \boldsymbol{\Sigma}_i^{-1} \boldsymbol{D}_i)$ is the OPG.
5. *The Fisher information matrix $\boldsymbol{A}(\boldsymbol{\beta}_0)$ and the OPG $\boldsymbol{B}(\boldsymbol{\beta}_0)$ can be strongly consistently estimated by*
$$\hat{\boldsymbol{A}}(\hat{\boldsymbol{\beta}}) = \frac{1}{n} \sum_{i=1}^{n} \hat{\boldsymbol{A}}_i = \frac{1}{n} \sum_{i=1}^{n} \Big(\hat{\boldsymbol{D}}_i' \hat{\boldsymbol{\Sigma}}_i^{-1} \hat{\boldsymbol{D}}_i\Big) \quad \text{and}$$
$$\hat{\boldsymbol{B}}(\hat{\boldsymbol{\beta}}) = \frac{1}{n} \sum_{i=1}^{n} \hat{\boldsymbol{B}}_i = \frac{1}{n} \sum_{i=1}^{n} \Big(\hat{\boldsymbol{D}}_i' \hat{\boldsymbol{\Sigma}}_i^{-1} \hat{\boldsymbol{\Omega}}_i \hat{\boldsymbol{\Sigma}}_i^{-1} \hat{\boldsymbol{D}}_i\Big) \,,$$
where $\hat{\boldsymbol{D}}_i$, $\hat{\boldsymbol{\mu}}_i = \boldsymbol{\mu}(\boldsymbol{X}_i, \hat{\boldsymbol{\beta}})$, and $\hat{\boldsymbol{\Sigma}}_i = \boldsymbol{\Sigma}(\boldsymbol{X}_i, \hat{\boldsymbol{\beta}}, \boldsymbol{\Psi}_i)$ are the estimators of $\boldsymbol{D}_i$, $\boldsymbol{\mu}_i$, and $\boldsymbol{\Sigma}_i$, respectively, and $\hat{\boldsymbol{\Omega}}_i = \big(\boldsymbol{y}_i - \hat{\boldsymbol{\mu}}_i\big)\big(\boldsymbol{y}_i - \hat{\boldsymbol{\mu}}_i\big)'$.
6. *Necessary for the strong consistency of a PML1 estimator associated with a family of assumed distributions $f(\boldsymbol{y}_i | \boldsymbol{X}_i || \boldsymbol{\beta})$ for any parameter space, parameter vector $\boldsymbol{\beta}$, mean structure, and true distribution $f^\star$ is that the assumed distribution belongs to the linear exponential family.*
7. *The set of asymptotic covariance matrices of the PML1 estimator $\hat{\boldsymbol{\beta}}$ of $\boldsymbol{\beta}$ based on a linear exponential family has lower bound $\boldsymbol{\Upsilon}^{-1}(\boldsymbol{\beta}) = \big(\mathbb{E}^{\boldsymbol{X}}(\boldsymbol{D}_i' \boldsymbol{\Omega}_i^{-1} \boldsymbol{D}_i)\big)^{-1}$.*

Remark 5.3.

- For the strong consistency of the PML1 estimator, the true and the assumed density need not be equal. As a result, PML1 estimators are consistent but generally not efficient, if only the mean structure is correctly specified. The result on consistency holds true even if no further assumptions on the true conditional variance matrix $\boldsymbol{\Omega}(\boldsymbol{X}_i)$ or any other properties of the true distribution are introduced.
- The PML1 estimator is asymptotically normally distributed but the consistency of the estimated asymptotic covariance matrix of $\hat{\boldsymbol{\beta}}$ can be guaranteed only if the robust covariance matrix is used. Neither the OPG nor the Fisher information matrix is sufficient for consistent estimation of the asymptotic covariance matrix of $\hat{\boldsymbol{\beta}}$.
- Properties 5. and 6. of the theorem impose that the PML1 estimator is asymptotically equivalent to the ML estimator if the true family of distributions belongs to the linear exponential family and if the covariance matrix of the assumed distribution is correctly specified. Because the ML estimator is the best asymptotically unbiased and normally distributed (BAN) estimator, the PML1 estimator also is BAN in this special case.
- PML1 estimators are asymptotically efficient if the conditional covariance matrix of the assumed distribution $\boldsymbol{\Sigma}(\boldsymbol{X}_i, \boldsymbol{\beta}, \boldsymbol{\Psi}_i)$ equals the true conditional covariance matrix $\boldsymbol{\Omega}(\boldsymbol{X}_i)$. In this case, the covariance matrix of the true distribution has to be partially parameterized in $\boldsymbol{\beta}_0$, if the assumed conditional covariance matrix $\boldsymbol{\Sigma}(\boldsymbol{X}_i, \boldsymbol{\beta}, \boldsymbol{\Psi}_i)$ is partially parameterized in $\boldsymbol{\beta}_0$.
- The covariance matrix of Eq. 5.3 is termed a sandwich covariance matrix or robust covariance matrix, because it is robust to model misspecification in the sense discussed above. Its estimator is also sometimes called an empirical-based covariance estimator or Huber estimator. In contrast, the Fisher information matrix is often termed a model-based covariance matrix.
- The ideas underlying the proof to Property 6. of the theorem can be used to show that $\boldsymbol{C}(\boldsymbol{\beta}_0) - \boldsymbol{A}(\boldsymbol{\beta}_0)$ is positive semidefinite. However, in applications, it is conceivable that robust standard errors of single components of the parameter vector $\boldsymbol{\beta}$ based on $\boldsymbol{C}(\boldsymbol{\beta}_0)$ are smaller than model-based standard errors using $\boldsymbol{A}(\boldsymbol{\beta}_0)$.
- In most applications, the assumed covariance matrix $\boldsymbol{\Sigma}_i$ is estimated during the estimation process of $\boldsymbol{\beta}$ because it generally depends on $\boldsymbol{\beta}$. Note that PML1 estimation does not allow estimation of the nuisance parameter $\boldsymbol{\alpha}$. This generalization of the PML approach is considered in the next chapter.
- In general, $\ddot{\boldsymbol{\Omega}}_i = (\boldsymbol{y}_i - \hat{\boldsymbol{\mu}}_i)(\boldsymbol{y}_i - \hat{\boldsymbol{\mu}}_i)'$ is only a replacement but not a consistent estimator of $\boldsymbol{\Omega}(\boldsymbol{X}_i)$. The important result therefore is that the entire

expression $\boldsymbol{B}(\boldsymbol{\beta})$ can be consistently estimated although its component $\boldsymbol{\Omega}$ generally cannot be consistently estimated.

- The PML1 estimator is often obtained by a Fisher scoring algorithm using the Fisher information matrix of the assumed distribution. Because the algorithm is based on the assumed distribution, and not on the true distribution, it is termed modified Fisher scoring.
- The estimator of the OPG is biased (see Theorem 5.4), and several improvements to the estimator have been proposed (Sect. 5.4.2).

Proof.
1.: The proof is identical to the proof on existence of the ML estimator, and the reader may therefore refer to the literature (see, e.g., White, 1982, Theorem 2.1).
2.: A detailed proof of the strong consistency can be found, e.g., in Gourieroux and Monfort (1995a, p. 239). To prove the strong consistency, we have to show that the expected value of the kernel of the loglikelihood has a unique maximum at $\boldsymbol{\beta}_0$. This is true because

$$\begin{aligned}\mathbb{E}^{\boldsymbol{X}}\mathbb{E}^{\boldsymbol{y}}_{\boldsymbol{\beta}_0}\big(l(\boldsymbol{\beta})\big) &= a\big(\boldsymbol{\mu}_{i0},\boldsymbol{\Psi}_i\big) + \boldsymbol{c}\big(\boldsymbol{\mu}_{i0},\boldsymbol{\Psi}_i\big)'\mathbb{E}^{\boldsymbol{X}}\mathbb{E}^{\boldsymbol{y}}_{\boldsymbol{\beta}_0}(\boldsymbol{y}_i)\\ &= a\big(\boldsymbol{\mu}_{i0},\boldsymbol{\Psi}_i\big) + \boldsymbol{c}\big(\boldsymbol{\mu}_{i0},\boldsymbol{\Psi}_i\big)'\boldsymbol{\mu}_{i0}\,,\end{aligned}$$

where $\boldsymbol{\mu}_{i0} = \boldsymbol{\mu}(\boldsymbol{X}_i,\boldsymbol{\beta}_0)$. The result now follows from Property 1.6 and the first-order identifiability of $\boldsymbol{\beta}$.
3.: The score vector is derived as a byproduct below.
4.: The proof follows the same lines as the proof of statement 3. from Theorem 4.2. We therefore derive only the OPG and the Fisher information matrix. In the following, we omit the index i for simplicity, and we use the abbreviations $a = a(\boldsymbol{\mu},\boldsymbol{\Psi})$ and $\boldsymbol{c} = \boldsymbol{c}(\boldsymbol{\mu},\boldsymbol{\Psi})$. The individual score vector is given by

$$\boldsymbol{u}(\boldsymbol{\beta}) = \frac{\partial}{\partial\boldsymbol{\beta}}(a+\boldsymbol{c}'\boldsymbol{y}) \overset{(\star)}{=} \frac{\partial\boldsymbol{\mu}'}{\partial\boldsymbol{\beta}}\left(\frac{\partial a}{\partial\boldsymbol{\mu}} + \frac{\partial\boldsymbol{c}'}{\partial\boldsymbol{\mu}}\boldsymbol{y}\right) \overset{(\star\star)}{=} \frac{\partial\boldsymbol{\mu}'}{\partial\boldsymbol{\beta}}\frac{\partial\boldsymbol{c}'}{\partial\boldsymbol{\mu}}(\boldsymbol{y}-\boldsymbol{\mu}) = \boldsymbol{D}'\boldsymbol{\Sigma}^{-1}(\boldsymbol{y}-\boldsymbol{\mu})\,,$$

with the chain rule being applied at $(\star)$ and Property 1.4 at $(\star\star)$. The overall score vector is then obtained by summation and multiplication with $n-1$.

Subsequently, the OPG $\boldsymbol{B}(\boldsymbol{\beta}_0)$ can be derived as

$$\begin{aligned}\boldsymbol{B}(\boldsymbol{\beta}_0) &= \mathbb{E}^{\boldsymbol{X}}\mathbb{E}^{\boldsymbol{y}}\Big(\big(\boldsymbol{u}(\boldsymbol{\beta}_0)\boldsymbol{u}(\boldsymbol{\beta}_0)'\big)\Big) = \mathbb{E}^{\boldsymbol{X}}\left(\frac{\partial\boldsymbol{\mu}'}{\partial\boldsymbol{\beta}}\frac{\partial\boldsymbol{c}'}{\partial\boldsymbol{\mu}}\Big(\mathbb{E}^{\boldsymbol{y}}(\boldsymbol{y}-\boldsymbol{\mu})(\boldsymbol{y}-\boldsymbol{\mu})'\Big)\frac{\partial\boldsymbol{c}}{\partial\boldsymbol{\mu}'}\frac{\partial\boldsymbol{\mu}}{\partial\boldsymbol{\beta}'}\right)\\ &= \mathbb{E}^{\boldsymbol{X}}\big(\boldsymbol{D}'\boldsymbol{\Sigma}^{-1}\boldsymbol{\Omega}(\boldsymbol{X})\boldsymbol{\Sigma}^{-1}\boldsymbol{D}\big)\,.\end{aligned}$$

The Fisher information matrix is obtained as follows:

$$
\begin{aligned}
\boldsymbol{W}(\boldsymbol{\beta}) &= \frac{\partial^2}{\partial\boldsymbol{\beta}\partial\boldsymbol{\beta}'}(a+\boldsymbol{c}'\boldsymbol{y}) \\
&= \frac{\partial}{\partial\boldsymbol{\beta}}\left(\frac{\partial(a+\boldsymbol{c}'\boldsymbol{y})}{\partial\boldsymbol{\mu}'}\frac{\partial\boldsymbol{\mu}}{\partial\boldsymbol{\beta}'}\right) = \frac{\partial}{\partial\boldsymbol{\beta}}\left(\sum_{l=1}^{q}\frac{\partial\mu_l}{\partial\boldsymbol{\beta}'}\left(\frac{\partial a}{\partial\mu_l}+\frac{\partial\boldsymbol{c}'}{\partial\mu_l}\boldsymbol{y}\right)\right) \\
(\star) &= \sum_{l=1}^{q}\frac{\partial^2\mu_l}{\partial\boldsymbol{\beta}\partial\boldsymbol{\beta}'}\left(\frac{\partial a}{\partial\mu_l}+\frac{\partial\boldsymbol{c}'}{\partial\mu_l}\boldsymbol{y}\right) \\
&\quad+\sum_{j=1}^{q}\sum_{l=1}^{q}\left(\frac{\partial\mu_j\partial\mu_l}{\partial\boldsymbol{\beta}\partial\boldsymbol{\beta}'}\frac{\partial^2 a}{\partial\mu_j\partial\mu_l}+\frac{\partial\mu_j\partial\mu_l}{\partial\boldsymbol{\beta}\partial\boldsymbol{\beta}'}\frac{\partial^2\boldsymbol{c}'\boldsymbol{y}}{\partial\mu_j\partial\mu_l}\right) \\
(\star\star) &= \sum_{l=1}^{q}\frac{\partial^2\mu_l}{\partial\boldsymbol{\beta}\partial\boldsymbol{\beta}'}\frac{\partial\boldsymbol{c}'}{\partial\mu_l}(\boldsymbol{y}-\boldsymbol{\mu})+\frac{\partial\boldsymbol{\mu}'}{\partial\boldsymbol{\beta}}\left(\frac{\partial^2 a}{\partial\boldsymbol{\mu}\partial\boldsymbol{\mu}'}+\sum_{l=1}^{q}\frac{\partial^2 c_l}{\partial\boldsymbol{\mu}\partial\boldsymbol{\mu}'}y_l\right)\frac{\partial\boldsymbol{\mu}}{\partial\boldsymbol{\beta}'} \\
(\star\star\star) &= \sum_{l=1}^{q}\frac{\partial^2\mu_l}{\partial\boldsymbol{\beta}\partial\boldsymbol{\beta}'}\frac{\partial\boldsymbol{c}'}{\partial\mu_l}(\boldsymbol{y}-\boldsymbol{\mu})+\frac{\partial\boldsymbol{\mu}'}{\partial\boldsymbol{\beta}}\left(\sum_{l=1}^{q}\frac{\partial^2 c_l}{\partial\boldsymbol{\mu}\partial\boldsymbol{\mu}'}(y_l-\mu_l)-\frac{\partial\boldsymbol{c}'}{\partial\boldsymbol{\mu}}\right)\frac{\partial\boldsymbol{\mu}}{\partial\boldsymbol{\beta}'}\,.
\end{aligned}
$$

At $(\star)$ we use the chain rule, at $(\star\star)$ Property 1.4 and simple matrix manipulations. Finally, $(\star\star\star)$ results from Property 1.5.

The expected value of the first two terms of the last equation are $\mathbf{0}$ because $\mathbb{E}^{\boldsymbol{y}}(\boldsymbol{y}-\boldsymbol{\mu})=\mathbf{0}$. The Fisher information matrix is therefore given by

$$
\boldsymbol{A}(\boldsymbol{\beta}_0)=\mathbb{E}^{\boldsymbol{X}}\mathbb{E}^{\boldsymbol{y}}_{f^\star}\big(-\boldsymbol{W}(\boldsymbol{\beta}_0)\big)=\mathbb{E}^{\boldsymbol{X}}\!\left(\frac{\partial\boldsymbol{\mu}'}{\partial\boldsymbol{\beta}}\frac{\partial\boldsymbol{c}'}{\partial\boldsymbol{\mu}}\frac{\partial\boldsymbol{\mu}}{\partial\boldsymbol{\beta}'}\right)=\mathbb{E}^{\boldsymbol{X}}\!\big(\boldsymbol{D}'\boldsymbol{\Sigma}^{-1}\boldsymbol{D}\big)\,.
$$

5.: See, e.g., White (1981, Theorem 3.3).
6.: See, e.g., Gourieroux et al. (1984b, Appendix 2).
7.: Let $\boldsymbol{\Sigma}^{1/2}$ denote a root of the covariance matrix $\boldsymbol{\Sigma}$ fulfilling $\boldsymbol{\Sigma}_i=\boldsymbol{\Sigma}_i^{1/2}\boldsymbol{\Sigma}_i^{1/2'}$. It can be obtained, e.g., by performing an eigendecomposition of $\boldsymbol{\Sigma}$ and subsequent multiplication of the square root of the diagonal matrix of eigenvalues with the matrix of eigenvectors. Alternative square roots can be obtained from appropriate decompositions, such as the Cholesky decomposition.

The lower bound of the PML1 estimator is $\boldsymbol{\Upsilon}^{-1}$ because

$$
\begin{aligned}
\boldsymbol{A}^{-1}\boldsymbol{B}\boldsymbol{A}^{-1}-\boldsymbol{\Upsilon}^{-1}=\mathbb{E}^{\boldsymbol{X}}\Big(&\big(\boldsymbol{\Upsilon}\boldsymbol{D}'\boldsymbol{\Omega}^{-1}\boldsymbol{\Sigma}^{1/2}-\boldsymbol{A}^{-1}\boldsymbol{D}'\boldsymbol{\Sigma}^{-1/2}\big)\\
&\boldsymbol{\Sigma}^{-1/2}\boldsymbol{\Omega}\boldsymbol{\Sigma}^{-1/2}\\
&\big(\boldsymbol{\Sigma}^{1/2}\boldsymbol{\Omega}^{-1}\boldsymbol{D}\boldsymbol{\Upsilon}-\boldsymbol{\Sigma}^{-1/2}\boldsymbol{D}\boldsymbol{A}^{-1}\big)\Big)\geq\mathbf{0}\,.
\end{aligned}
$$

□

In the following theorem, we show that $\hat{\boldsymbol{B}}_i$ is a biased estimator of the OPG. This theorem also forms one basis for the bias corrections and small sample improvements of the robust covariance matrix, which are discussed in Sect. 5.4.2.

Theorem 5.4.

1. *The expected valued of the individual OPG* $\mathbb{E}^{\boldsymbol{X}}\mathbb{E}^{\boldsymbol{y}}(\hat{\boldsymbol{B}}_i) = \mathbb{E}^{\boldsymbol{X}}\mathbb{E}^{\boldsymbol{y}}(\hat{\boldsymbol{u}}_i\hat{\boldsymbol{u}}_i')$ *can be approximated by*

$$\mathbb{E}^{\boldsymbol{X}}\mathbb{E}^{\boldsymbol{y}}(\hat{\boldsymbol{u}}_i\hat{\boldsymbol{u}}_i') \approx (\boldsymbol{I} - \boldsymbol{A}_i\boldsymbol{A}^{-1})\boldsymbol{B}_i(\boldsymbol{I} - \boldsymbol{A}_i\boldsymbol{A}^{-1})' + \sum_{j\neq i}\boldsymbol{A}_i'\boldsymbol{A}^{-1}\boldsymbol{B}_j\boldsymbol{A}^{-1}\boldsymbol{A}_i\,, \tag{5.4}$$

where $\boldsymbol{A}_i = \boldsymbol{A}_i(\boldsymbol{\beta}_0) = \boldsymbol{D}_i'\boldsymbol{\Sigma}_i^{-1}\boldsymbol{D}_i$.

2. *The expected value of the outer product of the individual estimated ordinary residual can be approximated by*

$$\mathbb{E}^{\boldsymbol{X}}\mathbb{E}^{\boldsymbol{y}}(\hat{\varepsilon}_i\hat{\varepsilon}_i') \approx (\boldsymbol{I} - \boldsymbol{H}_{ii})\boldsymbol{\Omega}_i(\boldsymbol{I} - \boldsymbol{H}_{ii})' + \sum_{j\neq i}\boldsymbol{H}_{ij}\boldsymbol{\Omega}_j\boldsymbol{H}_{ij}'\,, \tag{5.5}$$

where $\boldsymbol{H}_{ij} = \boldsymbol{D}_i\boldsymbol{A}^{-1}\boldsymbol{D}_j'\boldsymbol{\Sigma}_j^{-1}$.

Remark 5.5.

- The first part of Theorem 5.4 provides an approximation to the OPG, while Eq. 5.5 approximates the inner term of the OPG.
- Part 2. of the theorem shows that $\mathbb{E}^{\boldsymbol{X}}\mathbb{E}^{\boldsymbol{y}}(\hat{\varepsilon}_i\hat{\varepsilon}_i')$ differs from $\mathbb{V}\text{ar}(\boldsymbol{y}_i) = \boldsymbol{\Omega}_i$. The robust variance estimator thus is biased.
- Both approximations can be used for bias corrections, which will be considered in detail in Sect. 5.4.2. To give an example, Mancl and DeRouen (2001) assumed that the last term of Eq. 5.5 is negligible. They argued that all elements of $\boldsymbol{H}_{ij}$ are between 0 and 1, and they are usually close to 0. Thus, "it may be reasonable to assume that the summation makes only a small contribution to the bias." This argument is substantiated by noting that $\boldsymbol{H}_{ii}$ is closely related to the leverage of subject i (see, e.g., Preisser and Qaqish, 1996; Ziegler and Arminger, 1996). Subsequently, they proposed to replace $\hat{\varepsilon}_i$ in the OPG with $\hat{\hat{\varepsilon}}_i = \hat{\varepsilon}_i/(1-h_{ii})^{1/2}$, where $h_{ij} = [\boldsymbol{X}(\boldsymbol{X}'\boldsymbol{X})^{-1}\boldsymbol{X}']_{ij}$ denotes the ijth element of the hat matrix $\boldsymbol{H}$.

Proof.
1.: For simplicity, we ignore the expectation over $\boldsymbol{X}$ in the proof. The approximation to the bias can be derived from a first order Taylor series of the individual score vector $\boldsymbol{u}_i(\hat{\boldsymbol{\beta}})$ around $\boldsymbol{\beta}_0$:

$$\hat{\boldsymbol{u}}_i(\hat{\boldsymbol{\beta}}) \overset{a.s.}{=} \boldsymbol{u}_i(\boldsymbol{\beta}_0) + \boldsymbol{W}_i(\boldsymbol{\beta}^*)(\hat{\boldsymbol{\beta}} - \boldsymbol{\beta}_0)\,, \tag{5.6}$$

for $|\boldsymbol{\beta}^* - \boldsymbol{\beta}_0| \le |\hat{\boldsymbol{\beta}} - \boldsymbol{\beta}_0|$. This Taylor series is analogous to the Taylor series of Eq. 4.5, which was used in the proof to Theorem 4.2. $\boldsymbol{W}_i(\boldsymbol{\beta}^*)$ converges to $-\boldsymbol{A}_i(\boldsymbol{\beta})$. If we take the outer product of Eq. 5.6 and the expected value with respect to $\boldsymbol{y}$, we obtain

$$
\begin{aligned}
\mathbb{E}^{\boldsymbol{y}}(\hat{\boldsymbol{u}}_i\hat{\boldsymbol{u}}_i') \approx \mathbb{E}^{\boldsymbol{y}}(\boldsymbol{u}_i\boldsymbol{u}_i') - \mathbb{E}^{\boldsymbol{y}}\Big(\boldsymbol{u}_i(\hat{\boldsymbol{\beta}}-\boldsymbol{\beta}_0)'\Big)\boldsymbol{A}_i - \boldsymbol{A}_i\mathbb{E}^{\boldsymbol{y}}\Big((\hat{\boldsymbol{\beta}}-\boldsymbol{\beta}_0)\boldsymbol{u}_i'\Big)& \\
+\boldsymbol{A}_i\mathbb{E}^{\boldsymbol{y}}\Big((\hat{\boldsymbol{\beta}}-\boldsymbol{\beta}_0)(\hat{\boldsymbol{\beta}}-\boldsymbol{\beta}_0)'\Big)\boldsymbol{A}_i\,.& \qquad (5.7)
\end{aligned}
$$

Next, we use the first-order Taylor series expansion for $\hat{\boldsymbol{\beta}}-\boldsymbol{\beta}_0$ from Eq. 4.5, i.e., $(\hat{\boldsymbol{\beta}}-\boldsymbol{\beta}_0) \stackrel{a.s.}{=} \big(-\sum_{i=1}^n \boldsymbol{W}_i(\boldsymbol{\beta}^*)\big)^{-1}\Big(\sum_{i=1}^n \boldsymbol{u}_i(\boldsymbol{\beta}_0)\Big)$. As before, $-\sum_{i=1}^n \boldsymbol{W}_i(\boldsymbol{\beta}^*)$ is replaced by its expected value $\boldsymbol{A}(\boldsymbol{\beta}_0) = \boldsymbol{A}$, and we note that $\mathbb{E}^{\boldsymbol{y}}(\boldsymbol{u}_i\boldsymbol{u}_j') = \boldsymbol{0}$ for $i \neq j$. Therefore, we can rewrite Eq. 5.7 as

$$
\begin{aligned}
\mathbb{E}^{\boldsymbol{y}}(\hat{\boldsymbol{u}}_i\hat{\boldsymbol{u}}_i') &\approx \boldsymbol{B}_i - \mathbb{E}^{\boldsymbol{y}}\Big(\boldsymbol{u}_i\big(\textstyle\sum\boldsymbol{u}_i\big)'\Big)\boldsymbol{A}\boldsymbol{A}_i - \boldsymbol{A}_i\boldsymbol{A}\mathbb{E}^{\boldsymbol{y}}\Big(\big(\textstyle\sum\boldsymbol{u}_i\big)\boldsymbol{u}_i'\Big) \\
&\quad +\boldsymbol{A}_i\boldsymbol{A}\,\mathbb{E}^{\boldsymbol{y}}\Big(\big(\textstyle\sum\boldsymbol{u}_i\big)\big(\textstyle\sum\boldsymbol{u}_i\big)'\Big)\boldsymbol{A}\boldsymbol{A}_i \\
&= \boldsymbol{B}_i - \boldsymbol{B}_i\boldsymbol{A}\boldsymbol{A}_i - \boldsymbol{A}_i\boldsymbol{A}\boldsymbol{B}_i + \boldsymbol{A}_i\boldsymbol{A}\sum_{j=1}^n \boldsymbol{B}_j\boldsymbol{A}\boldsymbol{A}_i\,,
\end{aligned}
$$

which completes the proof.
2.: The proof is carried out analogously to the proof of 1. with the exception that a Taylor series expansion of $\hat{\boldsymbol{\varepsilon}}_i$ around $\boldsymbol{\beta}_0$ is used. □

5.3 Examples

In this section, we derive several PML1 estimators and their robust variance and covariance matrix, respectively. We begin with two simple models based on the Poisson and the binomial distribution, where no further covariates are available. Subsequently, the normal distribution is considered in various examples. It includes the two-sample situation with known variances. Third, we consider the linear regression model with heteroscedastic errors. This is followed by the logistic regression model with assumed variance 1. The last two examples are two specific GEE models. One is the IEE with the covariance matrix being identical to the identity matrix, and the final example is a GEE1 model with fixed covariance matrix.

5.3.1 Simple pseudo maximum likelihood 1 models

In this section, we consider two simple PML1 models without covariates. We start by estimating the mean from an assumed Poisson distribution, followed by the problem of estimating a proportion based on an assumed binomial distribution.

Example 5.6 (Poisson mean). Consider n independently and identically distributed positive integer valued random variables $y_1, \ldots, y_n$ with positive real valued mean $\lambda = \beta > 0$. The true distribution might be any distribution with mean $\lambda = \mu > 0$ and finite variance $\sigma^2 < \infty$. However, in many applications, the Poisson distribution is chosen for statistical testing or confidence interval estimation.

In terms of Sect. 5.1, this means that the assumed distribution is given by $f(y_i||\lambda) = \lambda_i^y e^{-\lambda}/y_i$. We furthermore assume that the mean is correctly specified, i.e., $\mathbb{E}_{f^\star}(y_i||\lambda) = \mathbb{E}_f(y_i||\lambda) = \lambda$. With an assumed Poisson distribution, we already have $\mathbb{V}\text{ar}(y_i) = \lambda$. This Poisson model need not be true. For example, there might be some over- or underdispersion so that the true variance could be $\mathbb{V}\text{ar}_{f^\star}(y_i) = \Phi\,\lambda$ for some $\Phi > 0$. However, $\mathbb{V}\text{ar}(y_i)$ might be any other function or value, possibly independent of λ.

If the Poisson model were correct, $\sqrt{n}\big(\hat{\lambda} - \lambda\big) = \sqrt{n}\big(\bar{y} - \lambda\big)$ would be asymptotically normal with mean 0 and variance given by the inverse Fisher information. The Fisher information can be deduced from the second derivative (see Example 4.19). With $A_i(\lambda) = 1/\lambda$, we obtain $\hat{A}_i(\hat{\lambda}) = 1/\hat{\lambda}$. $\widehat{\mathbb{V}\text{ar}}(\hat{\lambda})$ can therefore be estimated by $\hat{\lambda}/n$ so that an asymptotic confidence interval for λ at confidence level $1 - \alpha$ would be given by

$$\hat{\lambda} \pm z_{1-\alpha/2}\sqrt{\hat{\lambda}/n}\,.$$

Here, $z_{1-\alpha/2}$ denotes the $1 - \frac{\alpha}{2}$ quantile of the standard normal distribution function.

If the simple Poisson model has to be doubted, one might prefer using the assumed distribution f together with the robust variance rather than the model-based variance. Because $\hat{\boldsymbol{D}} = \boldsymbol{1}_n$, $\hat{\boldsymbol{\Sigma}}^{-1} = \text{diag}(1/\hat{\lambda})$ and $\hat{\boldsymbol{\Omega}} = \text{diag}(\hat{\varepsilon}_i^2) = \text{diag}\big((y_i - \hat{\lambda})^2\big)$, we obtain $\hat{B}(\hat{\lambda}) = \frac{1}{n}\sum_{i=1}^{n}\big((y_i - \hat{\lambda})^2/\hat{\lambda}^2\big)$. The robust variance estimator of $\sqrt{\hat{\lambda}}$ based on the assumed Poisson distribution f is therefore given by

$$\hat{C}(\hat{\lambda}) = \hat{A}^{-1}(\hat{\lambda})\,\hat{B}(\hat{\lambda})\,\hat{A}^{-1}(\hat{\lambda}) = \hat{\lambda}\left(\tfrac{1}{n}\sum\nolimits_{i=1}^{n}\tfrac{(y_i-\hat{\lambda})^2}{\hat{\lambda}^2}\right)\hat{\lambda} = \tfrac{1}{n}\sum\nolimits_{i=1}^{n}(y_i - \hat{\lambda})^2\,.$$

A robust $1 - \alpha/2$ confidence interval of $\hat{\lambda}$ based on the assumed Poisson distribution is

$$\hat{\lambda} \pm z_{1-\alpha/2}\sqrt{\hat{C}(\hat{\lambda})/n} = \hat{\lambda} \pm z_{1-\alpha/2}\sqrt{\textstyle\sum_{i=1}^{n}(y_i - \hat{\lambda})^2/n^2}\,.$$

Again, we want to stress that this confidence interval is asymptotically valid even if the true distribution is not Poisson. For example, the data might come from a binomial distribution. The point estimator $\hat{\pi} = \bar{x}$ is the appropriate PML1 estimator, and the robust variance estimator adequately adjusts for the misspecification of the variance.

If, however, the Poisson model is true, the asymptotic relative efficiency of the robust variance is reduced compared with the model-based variance, i.e., the Fisher information. This is discussed to some extent in the last section of this chapter.

Example 5.7 (Probability of a binomial model). Consider n dichotomous independently and identically distributed random variables y_i, and let $y_i \sim B(m, \pi)$ be the assumed distribution, where n and m are fixed. Both the ML and the PML1 estimators of $\pi = \beta$ are $\hat{\pi} = \bar{y}/m$. The estimators of the model-based and the robust variance of $\hat{\pi}$ are given by (see, e.g., Royall, 1986, p. 223)

$$\hat{A}(\hat{\pi}) = \frac{(\bar{y}/m)\left(1 - (\bar{y}/m)\right)}{mn} \quad \text{and} \quad \hat{C}(\hat{\pi}) = \frac{\sum_{i=1}^{n}(y_i - \bar{y})^2}{m^2 n^2}, \tag{5.8}$$

respectively, where subscripts are used to denote the estimators. In many applications, the binomial model is inadequate, and y_i might show some over- or underdispersion, in analogy to the Poisson model. Alternatively, the true distribution of y_i might be the hypergeometric distribution. The variance of the joint distribution of n hypergeometric distributed random variables is given by $\mathbb{V}\text{ar}_{\text{HG}}(\hat{\pi}) = \frac{\pi(1-\pi)}{m}\frac{N-m}{N-1}$. While the robust variance estimator is strongly consistent for $\mathbb{V}\text{ar}(\hat{\pi})$ thus converges to $\mathbb{V}\text{ar}_{\text{HG}}(\hat{\pi})$ as n tends to ∞, the ML variance estimator, i.e., the Fisher information, does not.

5.3.2 Linear regression with heteroscedasticity

In this section, PML1 estimation of the mean based on an assumed normal distribution is considered in several examples. In the first example, we assume the variance to be equal to 1, and in the second, it is assumed to be fixed and equal to σ^2. The examples illustrate that PML1 estimation can be employed even if the assumed distribution obviously is misspecified. The sandwich estimator maintains consistent but generally not efficient estimation of the variance. The third example considers estimation of the mean from a normal distribution with two random variances. This example illustrates differences between different variance estimators, especially between those based on the expected and observed Fisher information and the robust variance estimator.

In the final example, we estimate the linear regression from an assumed homoscedastic regression model with fixed variance. With the PML1 method, consistent estimation of the parameter vector $\boldsymbol{\beta}$ and its variance is feasible even if the errors are heteroscedastic. One limitation of PML1 estimation is that it does not allow estimating a nuisance parameter from an assumed distribution, such as the normal distribution. The nuisance parameter is assumed to be known instead. A generalization to estimating the nuisance parameter in addition to the parameter of interest requires the quasi generalized PML (QGPML) method (Chapt. 6).

Example 5.8 (The simple mean model with variance **1***).* Consider the simple linear regression model $y_i = \mu + \varepsilon_i$, where the errors ε_i are independently and identically normally distributed. For the assumed distribution, we use $\sigma^2 = 1$.

The PML1 estimator is identical to the OLS estimator and given by $\hat{\mu} = \bar{x}$. The individual Fisher information is $A_i(\mu) = 1$, thus $\hat{A} = \hat{A}(\hat{\mu}) = 1$. With $\hat{\boldsymbol{D}} = \mathbf{1}_n$, $\hat{\boldsymbol{\Sigma}} = \text{diag}(1)$, and $\hat{\boldsymbol{\Omega}} = \text{diag}(\hat{\varepsilon}_i^2) = \text{diag}\big((y_i - \hat{\mu})^2\big)$, we obtain $\hat{B}(\hat{\mu}) = \frac{1}{n}\sum_{i=1}^n (y_i - \hat{\mu})^2$, and the sandwich estimator of $\sqrt{n}\,\hat{\mu}$ becomes $\hat{C}(\hat{\mu}) = \sum_{i=1}^n (y_i - \hat{\mu})^2/n$. Subsequently, the robust variance of $\hat{\mu}$ is given by $\sum_{i=1}^n \left(y_i - \hat{\mu}\right)^2/n^2$.

Example 5.9 (Simple mean model with variance $\boldsymbol{\sigma^2}$*).* Consider the same model as in Example 5.8 with the difference that the errors ε_i of the assumed distribution are now assumed to be independently and identically normally distributed with variance σ^2. Because $A_i(\mu) = 1/\sigma^2$ and $\hat{\boldsymbol{\Sigma}} = \text{diag}(\sigma^2)$ cancel out each other in the calculation of the robust variance, the robust variance estimator is identical to the one from the previous example.

Example 5.10 (Simple mean model with subject specific variances $\boldsymbol{\sigma_i^2}$*).* Consider the same model as in the two previous examples 5.8 with the difference that the errors ε_i of the assumed distribution are assumed to be independently and identically normally distributed with known variances σ_i^2, $\sigma_i^2 \neq \sigma_j^2$ for at least one pair (i, j) of indices $i, j = 1, \ldots, n$, $i \neq j$. The variances therefore are different for at least two subjects.

The PML1 estimator $\hat{\mu}$ for μ is $\hat{\mu} = \bar{y}$. The individual Fisher information is given by $A_i(\mu) = 1/\sigma_i^2$, and we therefore obtain $\hat{A}(\hat{\mu}) = \frac{1}{n}\sum_{i=1}^n \sigma_i^{-2}$. Furthermore, we derive $\hat{\boldsymbol{D}} = \mathbf{1}_n$, $\hat{\boldsymbol{\Sigma}} = \text{diag}(\sigma_i^2)$, and $\hat{\boldsymbol{\Omega}} = \text{diag}(\hat{\varepsilon}_i^2) = \text{diag}\big((y_i - \hat{\mu})^2\big)$ so that $\hat{B}(\hat{\mu}) = \frac{1}{n}\sum_{i=1}^n \left((y_i - \hat{\mu})^2/\sigma_i^4\right)$. The robust variance estimator of $\sqrt{n}\,\hat{\mu}$ is therefore given by $\hat{C}(\hat{\mu}) = n\big(\sum_{i=1}^n (y_i - \hat{\mu})^2/\sigma_i^4\big)/\big(\sum_{i=1}^n \sigma_i^{-2}\big)^2$. In this example, subject specific variances do not cancel out, and the robust variance therefore differs from the robust variance in the two previous examples.

Example 5.11 (Normal mean with random variance). The differences between variance estimates using the observed Fisher information matrix, the expected Fisher information matrix, and the robust variance matrix are illustrated in this example. We aim at estimating the general mean μ and use the normal distribution with variances determined by a fair coin toss as assumed distribution. If tail appears on toss number i, $y_i \sim N(\mu, \sigma_1^2)$. If head appears, then $y_i \sim N(\mu, \sigma_2^2)$. The variances σ_1^2 and σ_2^2 are assumed to be known with $\sigma_1^2 \neq \sigma_2^2$. Furthermore, we assume that this experiment leads to n_1 observations related to distribution 1 and n_2 observations related to 2, and we let $n = n_1 + n_2$.

The kernel of the joint normed pseudo loglikelihood is proportional to

$$l(\mu) \propto -\frac{1}{2}\sum_{i=1}^{n_1}\frac{(y_i-\mu)^2}{\sigma_1^2} - \frac{1}{2}\sum_{i=1}^{n_2}\frac{(y_i-\mu)^2}{\sigma_2^2}.$$

The ML estimator for μ is obtained as

$$\hat{\mu} = \frac{(n_1\bar{y}_1/\sigma_1^2) + (n_2\bar{y}_2/\sigma_2^2)}{(n_1/\sigma_1^2) + (n_2/\sigma_2^2)},$$

and the second derivative of the loglikelihood with respect to (w.r.t.) μ is given by

$$W(\mu) = -\frac{n_1}{\sigma_1^2} - \frac{n_2}{\sigma_2^2}. \tag{5.9}$$

Because the coin tossing is fair, the Fisher information is

$$A(\mu) = \frac{n/2}{\sigma_1^2} + \frac{n/2}{\sigma_2^2}, \tag{5.10}$$

and we similarly obtain

$$B(\mu) = \frac{1}{n}\left(\frac{\sum_{i=1}^{n_1}(y_1-\mu)^2}{\sigma_1^4} + \frac{\sum_{i=1}^{n_2}(y_1-\mu)^2}{\sigma_2^4}\right) \tag{5.11}$$

as OPG. Equation 5.9, 5.10, and 5.11 can be used to obtain three different estimators for $\mathbb{V}\text{ar}(\hat{\mu})$. If the expected Fisher information is used, the asymptotic variance of $\hat{\mu}$ is

$$2\frac{\sigma_1^2\sigma_2^2}{\sigma_1^2+\sigma_2^2}.$$

The variance based on the observed Fisher information is given by

$$n\frac{\sigma_1^2\sigma_2^2}{n_1\sigma_1^2+n_2\sigma_2^2},$$

and, finally, the robust variance estimator of $\hat{\mu}$ is

$$\frac{1}{n}\left(\frac{n_1}{\sigma_1^2}+\frac{n_2}{\sigma_2^2}\right)^2\left(\frac{\sum_{i=1}^{n_1}(y_i-\hat{\mu})^2}{\sigma_1^4} + \frac{\sum_{i=1}^{n_2}(y_i-\hat{\mu})^2}{\sigma_2^4}\right).$$

The expected Fisher information ignores the actual number of observations from the two different distributions, thus the possible imbalance. In contrast, the observed Fisher information is properly conditioned and takes into account the imbalance (Efron and Hinkley, 1978). Finally, the robust variance estimator is properly conditioned and also protects against possible errors in the assumed variances σ_1^2 and σ_2^2 (Royall, 1986).

Example 5.12 (Linear regression with heteroscedastic variance). Consider the classical multiple linear regression model with heteroscedasticity as the true model:

$$y_i = \boldsymbol{x}_i'\boldsymbol{\beta} + \varepsilon_i\,, \quad \mathbb{E}(\varepsilon_i|\boldsymbol{x}_i) = 0\,, \quad \mathbb{V}\mathrm{ar}(\varepsilon_i|\boldsymbol{x}_i) = \sigma_i^2\,, \quad i = 1, \dots, n\,,$$

where $\boldsymbol{x}_i, \boldsymbol{\beta} \in \mathbb{R}^p$. We collect elements in vectors and vectors in matrices: $\boldsymbol{y} = (y_1, \dots, y_n)'$, $\boldsymbol{\varepsilon} = (\varepsilon_1, \dots, \varepsilon_n)'$, and $\boldsymbol{X} = (\boldsymbol{x}_1, \dots, \boldsymbol{x}_n)'$.

For estimation, we choose as assumed distribution for $\boldsymbol{y}_i$ given $\boldsymbol{x}_i$ the normal distribution with mean $\mu_i = \boldsymbol{x}_i'\boldsymbol{\beta}$ and variance $\sigma_i^2 = 1$. The density of the assumed distribution is $f(y_i|\boldsymbol{x}_i||\boldsymbol{\beta}, \Psi) = \varphi(y_i|\mu_i = \boldsymbol{x}_i'\boldsymbol{\beta}, \sigma^2 = 1)$, with φ denoting the density of the normal distribution. The kernel of the normed pseudo loglikelihood is given by

$$l(\boldsymbol{\beta}) = \frac{1}{n}\sum_{i=1}^{n} -\frac{1}{2}(y_i - \boldsymbol{x}_i'\boldsymbol{\beta})^2\,,$$

which needs to be maximized w.r.t. $\boldsymbol{\beta}$. This maximization problem is equivalent to the ordinary least squares (OLS) problem, and differentiation w.r.t. $\boldsymbol{\beta}$ yields

$$\frac{\partial \ln f(y_i|\boldsymbol{x}_i||\boldsymbol{\beta}, 1)}{\partial \boldsymbol{\beta}} = \boldsymbol{x}_i(y_i - \boldsymbol{x}_i'\boldsymbol{\beta}) \quad \text{and} \quad \frac{\partial^2 \ln f(y_i|\boldsymbol{x}_i||\boldsymbol{\beta}, 1)}{\partial \boldsymbol{\beta}\partial \boldsymbol{\beta}'} = -\boldsymbol{x}_i\boldsymbol{x}_i'\,.$$

The MLE are therefore given by

$$\tfrac{1}{n}\sum_{i=1}^{n} \boldsymbol{x}_i(y_i - \boldsymbol{x}_i'\hat{\boldsymbol{\beta}}) = \tfrac{1}{n}\boldsymbol{X}'(\boldsymbol{y} - \boldsymbol{X}\hat{\boldsymbol{\beta}}) = \boldsymbol{0}\,,$$

the PML1 estimator is

$$\hat{\boldsymbol{\beta}} = \left(\boldsymbol{X}'\boldsymbol{X}\right)^{-1}\boldsymbol{X}'\boldsymbol{y}\,,$$

and the robust variance $\mathbb{V}\mathrm{ar}\big(\sqrt{n}\hat{\boldsymbol{\beta}}\big)$ of $\sqrt{n}\hat{\boldsymbol{\beta}}$ can be estimated consistently by

$$\widehat{\mathbb{V}\mathrm{ar}}\big(\sqrt{n}\hat{\boldsymbol{\beta}}\big) = \big(\tfrac{1}{n}\boldsymbol{X}'\boldsymbol{X}\big)^{-1}\big(\tfrac{1}{n}\boldsymbol{X}'\boldsymbol{D}\boldsymbol{X}\big)\big(\tfrac{1}{n}\boldsymbol{X}'\boldsymbol{X}\big)^{-1}\,,$$

where $\boldsymbol{D} = \mathrm{diag}(\hat{\varepsilon}_i^2)$. The robust variance $\mathbb{V}\mathrm{ar}\big(\hat{\boldsymbol{\beta}}\big)$ of $\hat{\boldsymbol{\beta}}$ is therefore given by $\big(\boldsymbol{X}'\boldsymbol{X}\big)^{-1}\big(\boldsymbol{X}'\boldsymbol{D}\boldsymbol{X}\big)\big(\boldsymbol{X}'\boldsymbol{X}\big)^{-1}$ and deviates from the usual model-based variance estimator $\sigma^2(\boldsymbol{X}'\boldsymbol{X})^{-1}$ for a common homoscedastic variance σ^2. This variance estimator is robust under any type of heteroscedasticity, and it is essentially the weighted jack-knife estimator of Hinkley (1977). An even stronger result can be deduced: The robust variance estimator for $\hat{\boldsymbol{\beta}}$ is asymptotically equivalent to a jack-knife estimator under any GEE model (Lipsitz et al., 1994a).

Further simple examples for deriving the robust variance estimator have been given by Royall (1986). Nice applications can be found in Binder (1983). He specifically derived the robust variance estimator for the coefficient of determination R^2.

5.3.3 Logistic regression with variance equal to 1

In this section, we consider two examples for logistic regression models. The first example is the standard logistic regression model, but we use the robust covariance matrix instead of the model-based covariance matrix. In the second example, we consider a model in which the true distribution is a logistic distribution, but we use the normal distribution with variance 1 as assumed distribution.

Example 5.13 (Logistic regression for dichotomous dependent variables with robust variance). We consider the univariate GLM of Sect. 3.1.3. We assume that y_i, $i = 1, \ldots, n$ are dichotomous independent random variables, and $\boldsymbol{x}_i$ is the vector of independent variables. The logit link is chosen as the link function so that

$$y_i = \mu(\boldsymbol{x}_i, \boldsymbol{\beta}) + \epsilon_i \quad \text{with} \quad \mu_i = \mu(\boldsymbol{x}_i, \boldsymbol{\beta}) = \frac{\exp(\eta_i)}{1+\exp(\eta_i)} = \frac{\exp(\boldsymbol{x}_i'\boldsymbol{\beta})}{1+\exp(\boldsymbol{x}_i'\boldsymbol{\beta})}$$

is the correctly specified mean structure, i.e., $\mathbb{E}_f(y_i) = \mathbb{E}_{f^\star}(y_i) = \mu_i$. The true covariance need not be specified $\mathbb{V}\mathrm{ar}_{f^\star}(y_i|\boldsymbol{x}_i) = \Omega(\boldsymbol{x}_i)$. We assume, however, that $y_i|\boldsymbol{x}_i$ is independent and distributed as $B(1, \mu_i)$. The resulting PML1 estimating equations are identical to the estimating equations from the standard univariate GLM with natural link function and given by (Sect. 4.4)

$$\mathbf{0} = \boldsymbol{u}(\hat{\boldsymbol{\beta}}) = \tfrac{1}{n}\boldsymbol{X}'(\boldsymbol{y} - \hat{\boldsymbol{\mu}})\,.$$

The robust covariance matrix of $\hat{\boldsymbol{\beta}}$ is given by

$$\widehat{\mathbb{V}\mathrm{ar}}(\hat{\boldsymbol{\beta}}) = \left(\boldsymbol{X}'\hat{\boldsymbol{\Sigma}}^{-1}\boldsymbol{X}\right)^{-1}\left(\boldsymbol{X}'\hat{\boldsymbol{\Omega}}\boldsymbol{X}\right)\left(\boldsymbol{X}'\hat{\boldsymbol{\Sigma}}^{-1}\boldsymbol{X}\right)^{-1} \tag{5.12}$$

with $\hat{\boldsymbol{\Sigma}} = \mathrm{diag}\big(\hat{\mu}_i(1-\hat{\mu}_i)\big)$ and $\hat{\boldsymbol{\Omega}} = \mathrm{diag}\big((y_i - \hat{\mu}_i)^2\big)$.

The robust covariance matrix in Eq. 5.12 has a different form from the robust covariance matrix of Theorem 5.2 because $\boldsymbol{\Sigma}^{-1}$ cancels out in the OPG. The reason is that $\partial l_i(\boldsymbol{\beta})/\partial\beta_j = (y_i - \mu_i)x_{ij}$ holds for natural link functions as shown in Eq. 4.17.

The interesting aspect of this example is that the robust covariance matrix yields consistent parameter estimates even if the variances are misspecified, e.g., when there is some over- or underdispersion.

Example 5.14 (Logistic regression for dichotomous dependent variables and variance **1***).* We consider the situation of the previous example, but we now use the normal distribution as assumed or "working" distribution. This assumption is undoubtedly incorrect if the dependent variable is dichotomous or – in our case – a sum of dichotomous random variables. With PML1 estimation, one can still obtain consistent parameter estimates for the mean structure and the variance of the parameter estimator if the mean structure is correctly specified.

Specifically, we assume that the assumed distribution is the normal distribution with mean $\mu_i = \mu(\boldsymbol{x}_i, \boldsymbol{\beta})$ and variance 1 so that the working density is given by

$$f(y_i|\boldsymbol{x}_i\,||\boldsymbol{\beta}, \Psi) = \varphi(y_i\,||\mu(\boldsymbol{x}_i, \boldsymbol{\beta}), \sigma^2 = 1)\,.$$

With $l(\boldsymbol{\beta}) = \frac{1}{n}\sum_{i=1}^n \big(-\frac{1}{2}(y_i - \mu_i)^2\big)$ being the kernel of the normed pseudo log likelihood, we obtain the PML1 estimating equations

$$\frac{1}{n}\sum_{i=1}^n \frac{\partial \hat{\mu}_i}{\partial \boldsymbol{\beta}}(y_i - \hat{\mu}_i) = \frac{1}{n}\sum_{i=1}^n \frac{e^{\hat{\eta}_i}}{(1+e^{\hat{\eta}_i})^2}\boldsymbol{x}_i\Big(y_i - \frac{e^{\hat{\eta}_i}}{1+e^{\hat{\eta}_i}}\Big) = \boldsymbol{0}\,,$$

which can be summarized to $\frac{1}{n}\hat{\boldsymbol{D}}'(\boldsymbol{y} - \hat{\boldsymbol{\mu}}) = \boldsymbol{0}$ in matrix notation with $\boldsymbol{D} = \partial\boldsymbol{\mu}/\partial\boldsymbol{\beta}' = \operatorname{diag}\big(e^{\eta_i}/(1+e^{\eta_i})^2\big)\boldsymbol{X}'$. The covariance matrix of $\hat{\boldsymbol{\beta}}$ can be estimated in this partly misspecified model by the robust covariance matrix

$$\widehat{\mathbb{V}\mathrm{ar}}(\hat{\boldsymbol{\beta}}) = \frac{1}{n}\hat{\boldsymbol{A}}^{-1}\hat{\boldsymbol{B}}\hat{\boldsymbol{A}}^{-1} = (\hat{\boldsymbol{D}}'\hat{\boldsymbol{D}})^{-1}(\hat{\boldsymbol{D}}'\hat{\boldsymbol{\Omega}}\hat{\boldsymbol{D}})^{-1}(\hat{\boldsymbol{D}}'\hat{\boldsymbol{D}})^{-1}$$

with $\hat{\boldsymbol{\Omega}} = \operatorname{diag}\big((y_i - \hat{\mu}_i)^2\big)$. The elements of $\hat{\boldsymbol{A}}$ and $\hat{\boldsymbol{B}}$ are given by

$$\hat{\boldsymbol{A}}_i = \left(\frac{e^{\hat{\eta}_i}}{(1+e^{\hat{\eta}_i})^2}\right)^2 \boldsymbol{x}_i\boldsymbol{x}_i' \quad\text{and}\quad \hat{\boldsymbol{B}}_i = \left(\frac{e^{\hat{\eta}_i}}{(1+e^{\hat{\eta}_i})^2}\right)^2 \boldsymbol{x}_i(y_i - \hat{\mu}_i)^2\boldsymbol{x}_i'\,.$$

5.3.4 Independence estimating equations with covariance matrix equal to identity matrix

In this example, we derive a special set of IEE using the PML1 method. These IEE are different from the commonly used IEE that were introduced by Liang and Zeger in a series of papers (see, e.g., Zeger et al., 1985; Liang and Zeger, 1986; Zeger and Liang, 1986; for a discussion, see Ziegler and Vens, 2010). The well-known IEE require the quasi generalized pseudo likelihood estimation approach which is discussed in the next chapter. Throughout this section we use the normal distribution as assumed distribution, and we assume that the mean structure is correctly specified. The identity matrix is specifically chosen as assumed covariance matrix. Again, although this estimation approach is not efficient in many instances, the parameter vector $\boldsymbol{\beta}$ and its variance can be consistently estimated.

We consider the T-dimensional random vector $\boldsymbol{y}_i = (y_{i1}, \dots, y_{iT})'$, and its associated matrix of explanatory variables $\boldsymbol{X}_i = (\boldsymbol{x}_{i1}, \dots, \boldsymbol{x}_{iT})'$, for $i = 1, \dots, n$. The pairs $(\boldsymbol{y}_i, \boldsymbol{X}_i)$ are assumed to be independent and identically distributed. The mean structure is given by

$$\mathbb{E}(\boldsymbol{y}_i|\boldsymbol{X}_i||\boldsymbol{\beta}_0) = g(\boldsymbol{X}_i\boldsymbol{\beta}_0)\,, \tag{5.13}$$

where the response function g is defined element-wise as in multivariate GLM. One important assumption of Eq. 5.13 is that the parameter vector $\boldsymbol{\beta}$ of interest is constant across t, a property termed time independence. The second relevant assumption is that the mean of $\boldsymbol{y}_i$ is correctly specified given the matrix of independent variables $\boldsymbol{X}_i$. This assumption is stricter than the assumption of standard GLM, where y_{it} is modeled as a function of $\boldsymbol{x}_{it}$ only. Equation 5.13 permits that elements of the vector of independent variables $\boldsymbol{x}_{it}$ observed at time t may have an effect, e.g., on $y_{i,t+1}$. It is fulfilled, e.g., if only independent variables are used that do not vary over time. It may, however, be violated in several applications, and a few biometrical examples are as follows (Ziegler and Vens, 2010):

- In family studies, the stress level of the parents may have an effect on the health status of the offspring. Similarly, the stress level of the offspring, e.g., because of examinations at school, may also affect the health status of their parents.
- In school studies, the exposure of other children may affect the health status of a particular child.
- In dental studies, the exposure level at a neighboring tooth may affect a particular tooth.

The consequences of this assumption are discussed in detail, e.g., in Pepe and Anderson (1994), Pan et al. (2002), and Schildcrout and Heagerty (2005).

No restrictions about the true covariance structure $\boldsymbol{\Omega}(\boldsymbol{X}_i)$ are imposed, and we only assume the existence of the true conditional variance matrix of $\boldsymbol{y}_i$ given $\boldsymbol{X}_i$ for all $i = 1, \dots, n$.

The assumed distribution of $\boldsymbol{y}_i$ given $\boldsymbol{X}_i$ is the normal distribution with mean structure $g(\boldsymbol{X}_i'\boldsymbol{\beta}_0)$ and covariance matrix $\boldsymbol{I} = \boldsymbol{I}_{T\times T}$, i.e., $\boldsymbol{y}_i|\boldsymbol{X}_i \sim N\big(g(\boldsymbol{X}_i\boldsymbol{\beta}_0), \boldsymbol{I}\big)$. Although this distributional assumption is most likely incorrect, we use it to illustrate the idea of PML1 estimation. The kernel of the individual pseudo loglikelihood function is given by

$$l_i(\boldsymbol{y}_i|\boldsymbol{X}_i||\boldsymbol{\beta}, \boldsymbol{I}) = -\tfrac{1}{2}\big(\boldsymbol{y}_i - g(\boldsymbol{X}_i\boldsymbol{\beta})\big)'\big(\boldsymbol{y}_i - g(\boldsymbol{X}_i\boldsymbol{\beta})\big)\,.$$

Differentiation with respect to $\boldsymbol{\beta}$ yields the score vector

$$\boldsymbol{u}(\boldsymbol{\beta}) = \frac{1}{n}\sum_{i=1}^{n} \boldsymbol{D}_i'\boldsymbol{\varepsilon}_i\,,$$

and the estimating equations

$$\boldsymbol{u}(\hat{\boldsymbol{\beta}}) = \frac{1}{n}\sum_{i=1}^{n} \hat{\boldsymbol{D}}_i' \hat{\boldsymbol{\varepsilon}}_i = \boldsymbol{0}\,,$$

where $\boldsymbol{D}_i = \partial\boldsymbol{\mu}_i/\partial\boldsymbol{\beta}'$ is the matrix of first derivatives, and $\boldsymbol{\varepsilon}_i = \boldsymbol{y}_i - \boldsymbol{\mu}_i = \boldsymbol{y}_i - g(\boldsymbol{X}_i\boldsymbol{\beta})$ is the first-order residual. These estimating equations are termed IEE with identity covariance matrix.

According to Theorem 5.2, the resulting estimator $\hat{\boldsymbol{\beta}}$ is asymptotically normally distributed. The Fisher information matrix $\boldsymbol{A}$ and the OPG $\boldsymbol{B}$ can be estimated (strongly) consistently by $\hat{\boldsymbol{A}}$ and $\hat{\boldsymbol{B}}$ with components $\hat{\boldsymbol{A}}_i = \hat{\boldsymbol{D}}_i'\hat{\boldsymbol{D}}_i$ and $\hat{\boldsymbol{B}} = \hat{\boldsymbol{D}}_i'\hat{\boldsymbol{\Omega}}_i\hat{\boldsymbol{D}}_i$, where $\hat{\boldsymbol{\Omega}}_i = \hat{\boldsymbol{\varepsilon}}_i\hat{\boldsymbol{\varepsilon}}_i'$ is the outer product of the estimated individual first order residuals.

Estimation using the identity matrix as assumed covariance matrix is inefficient in most applications because the identity matrix will be dissimilar to the true underlying covariance matrix. Estimation may therefore be improved by using a fixed covariance matrix that is "closer" to the true covariance matrix. This idea will be considered in the next section. However, instead of using a fixed covariance matrix, one might wish to estimate it from the data. In this case, the nuisance parameter from the linear exponential family has to be estimated, and it requires the quasi generalized pseudo maximum likelihood approach, which will be discussed in the next chapter.

5.3.5 Generalized estimating equations 1 with fixed covariance matrix

We consider the same model as in the previous section, but we use an arbitrary fixed covariance matrix $\boldsymbol{\Sigma}_i$ instead of the identity matrix. The starting point thus is the assumed distribution $\boldsymbol{y}_i|\boldsymbol{X}_i \sim N\big(g(\boldsymbol{X}_i\boldsymbol{\beta}_0), \boldsymbol{\Sigma}_i\big)$. Differentiation of the loglikelihood with respect to $\boldsymbol{\beta}$ yields the score vector

$$\boldsymbol{u}(\boldsymbol{\beta}) = \frac{1}{n}\sum_{i=1}^{n} \boldsymbol{D}_i'\boldsymbol{\Sigma}_i^{-1}\boldsymbol{\varepsilon}_i$$

and the estimating equations

$$\boldsymbol{u}(\hat{\boldsymbol{\beta}}) = \frac{1}{n}\sum_{i=1}^{n} \hat{\boldsymbol{D}}_i'\boldsymbol{\Sigma}_i^{-1}\hat{\boldsymbol{\varepsilon}}_i = \boldsymbol{0}\,.$$

The Fisher information matrix $\boldsymbol{A}$ and the OPG $\boldsymbol{B}$ can be (strongly) consistently estimated by

$$\hat{\boldsymbol{A}} = \frac{1}{n}\sum_{i=1}^{n} \hat{\boldsymbol{D}}_i' \boldsymbol{\Sigma}_i^{-1} \hat{\boldsymbol{D}}_i \quad \text{and} \quad \hat{\boldsymbol{B}} = \frac{1}{n}\sum_{i=1}^{n} \hat{\boldsymbol{D}}_i' \boldsymbol{\Sigma}_i^{-1} \hat{\boldsymbol{\Omega}}_i \boldsymbol{\Sigma}_i^{-1} \hat{\boldsymbol{D}}_i$$

with $\hat{\boldsymbol{\Omega}}_i = \hat{\varepsilon}_i \hat{\varepsilon}_i'$.

5.4 Efficiency and bias of the robust variance estimator

Several examples in the last section have illustrated the broad applicability of the PML1 estimation approach. However, PML1 estimation is not efficient as the classical ML approach if the assumed distribution equals the true distribution, i.e., if the model is correctly specified. The use of the robust covariance matrix will lead to a loss in efficiency. The efficiency of the PML1 approach will be discussed in the next section. Although the robust variance estimator yields consistent estimates under suitable regularity conditions if the mean structure is correctly specified, its small sample properties have been criticized. For example, the robust covariance estimator is biased in finite samples. Therefore, bias corrections and other approaches for improving the small sample properties of the robust covariance estimator are discussed in the final section of this chapter.

5.4.1 Efficiency considerations

PML1 estimation yields consistent but not necessarily efficient estimates. Basically, two questions arise: First, if we use the robust covariance matrix instead of the Fisher information matrix, is there always a loss in efficiency? Or are there situations in which the robust covariance matrix is more efficient than its model-based counterpart? Second, since there is no free lunch and since we lose efficiency for the extra robustness because we work with partly misspecified models, can we quantify the loss in efficiency? Analytical answers to both questions will be provided below. Although efficiency considerations are warranted, we want to stress that PML1 estimation yields consistent estimates even if second-order moments are misspecified. The ML approach requires, however, the correct model specification. Therefore, PML1 and ML can be compared with respect to efficiency only when the mean structure and the covariance matrix are correctly specified.

An answer to the first question, whether there always is a loss in efficiency, can be obtained by noting that the variance matrix of an asymptotically unbiased estimator is bounded by the Rao-Cramér bound, which is given by the inverse of the Fisher information matrix (see, e.g., Lehmann and Casella, 1998). We therefore do not expect that the robust variance matrix can be smaller than the model-based variance matrix, and this can also be shown without making use of the Rao-Cramér bound:

Theorem 5.15. *We consider the same situation as in Theorem 5.2. Then, the difference between the robust variance and the model-based variance matrix is non-negative, i.e.,* $\boldsymbol{C}(\boldsymbol{\beta}_0) - \boldsymbol{A}(\boldsymbol{\beta}_0) \geq \boldsymbol{0}$.

Thus, the robust variance is always larger than the model-based variance; thus, the robust variance cannot be more efficient.

Proof. The following elegant statistical proof of this statement is due to Dr. Rao Chaganty (personal communication). We first show that for a positive definite $nT \times nT$ matrix $\boldsymbol{\Omega}$ and an $nT \times p$ matrix $\boldsymbol{N}$ of rank p, $nT \geq p$, we have

$$\boldsymbol{\Omega} - \boldsymbol{N}(\boldsymbol{N}'\boldsymbol{\Omega}^{-1}\boldsymbol{N})^{-1}\boldsymbol{N}' \geq \boldsymbol{0}\,. \tag{5.14}$$

It is valid because we can verify $\mathbb{C}\text{ov}(\hat{\boldsymbol{y}}, \boldsymbol{y}) = \mathbb{V}\text{ar}(\hat{\boldsymbol{y}}) = \boldsymbol{N}(\boldsymbol{N}'\boldsymbol{\Omega}^{-1}\boldsymbol{N})^{-1}\boldsymbol{N}'$ for a random vector $\boldsymbol{y}$ with $\mathbb{V}\text{ar}(\boldsymbol{y}) = \boldsymbol{\Omega}$ and $\hat{\boldsymbol{y}} = \boldsymbol{N}(\boldsymbol{N}'\boldsymbol{\Omega}^{-1}\boldsymbol{N})^{-1}\boldsymbol{N}'\boldsymbol{\Omega}^{-1}\boldsymbol{y}$. Thus, $\mathbb{V}\text{ar}(\boldsymbol{y} - \hat{\boldsymbol{y}}) = \mathbb{V}\text{ar}(\boldsymbol{y}) - \mathbb{V}\text{ar}(\hat{\boldsymbol{y}}) = \boldsymbol{\Omega} - (\boldsymbol{N}'\boldsymbol{\Omega}^{-1}\boldsymbol{N})^{-1}\boldsymbol{N}'$.

In the next step, we pre- and post-multiply Eq. 5.14 by $nT \times p$ matrices $\boldsymbol{M}'$ and $\boldsymbol{M}$ of full rank p, yielding $\boldsymbol{M}'\boldsymbol{\Omega}\boldsymbol{M} - \boldsymbol{M}'\boldsymbol{N}(\boldsymbol{N}'\boldsymbol{\Omega}^{-1}\boldsymbol{N})^{-1}\boldsymbol{N}'\boldsymbol{M} \geq \boldsymbol{0}$. We now pre- and post-multiply with the inverses of $\boldsymbol{M}'\boldsymbol{N}$ and $\boldsymbol{N}'\boldsymbol{M}$, respectively, and obtain a matrix version of the Cauchy-Schwarz inequality

$$(\boldsymbol{M}'\boldsymbol{N})^{-1}\boldsymbol{M}'\boldsymbol{\Omega}\boldsymbol{M}(\boldsymbol{N}'\boldsymbol{M})^{-1} - (\boldsymbol{N}'\boldsymbol{\Omega}^{-1}\boldsymbol{N})^{-1} \geq \boldsymbol{0}\,.$$

The proof is completed by letting $\boldsymbol{M} = (\boldsymbol{M}_1', \dots, \boldsymbol{M}_n')'$ with $\boldsymbol{M}_i = \boldsymbol{\Sigma}_i^{-1}\boldsymbol{D}_i$, $\boldsymbol{N} = \boldsymbol{D} = (\boldsymbol{D}_1', \dots, \boldsymbol{D}_n')'$, and $\boldsymbol{\Omega} = \text{diag}(\boldsymbol{\Omega}_i)$, $i = 1, \dots, n$. □

The second question, which deals with the quantification of the loss in efficiency, has been considered in many different papers (Efron, 1986; Breslow, 1990; Firth, 1992; McCullagh, 1992; Carroll et al., 1998; Kauermann and Carroll, 2001). Unfortunately, the formulation and presentation of the general results are cumbersome, and the reader may refer to Kauermann and Carroll (2001) for these. However, it is possible to get an intuitive understanding on the possible loss of efficiency in simple models. We therefore first consider the simple Poisson model and the simple exponential model. Furthermore, we give the result for the heteroscedastic linear model and illustrate it using the design of a parallel group controlled clinical trial.

Theorem 5.16 (Efficiency of the sandwich estimator for a true Poisson model). *Consider a sample of n independently identically $Po(\lambda)$ distributed random variables $y_1, \dots, y_n$ with mean λ. The model-based variance of $\hat{\lambda}$ is estimated by $\hat{A}(\hat{\lambda}) = \bar{y}/n$, and the robust variance estimator is given by $\hat{C}(\hat{\lambda}) = s^2 = \frac{1}{n}\sum_{i=1}^{n}(y_i - \bar{y})^2$. If the Poisson model is true, the relative efficiency of the sandwich estimator, i.e., the ratio of the variances of the model-based and the robust variance estimator, is only*

$$\frac{\mathbb{V}ar(\hat{A})}{\mathbb{V}ar(\hat{C})} = \frac{n^2}{(n-1)^2} \frac{1}{1 + 2\frac{n}{n-1}\lambda},$$

and the asymptotic relative efficiency is

$$\lim_{n\to\infty} \frac{\mathbb{V}ar(\hat{A})}{\mathbb{V}ar(\hat{C})} = \frac{1}{1+2\lambda}.$$

For large λ, the asymptotic efficiency tends to 0 even if the sample size tends to infinity.

Proof. We first note that the variance $\mathbb{V}\text{ar}(s^2)$ of the sampling variance s^2 is given by (see, e.g., Kendall and Stuart, 1969, p. 244, Exercise 10.13)

$$\mathbb{V}\text{ar}(s^2) = \tfrac{(n-1)^2}{n^3}(\mu_4 - \mu_2^2) + \tfrac{2(n-1)}{n^3}\mu_2^2 = \tfrac{(n-1)^2}{n^3}\left(\mu_4 - \tfrac{n-3}{n-1}\mu_2^2\right), \qquad (5.15)$$

where μ_4 and $\mu_2 = \sigma^2$ denote the second and fourth central moments, respectively.

For the Poisson distribution, the fourth central moment μ_4 is given by $\lambda(1 + 3\lambda)$ (Kendall and Stuart, 1969, p. 89, Exercise 3.1), and $\sigma^4 = \lambda^2$. With $\mathbb{V}\text{ar}(\hat{A}) = \mathbb{V}\text{ar}(\bar{y}) = \lambda/n$ and $\mathbb{V}\text{ar}(\hat{C}) = \mathbb{V}\text{ar}(s^2) = \frac{(n-1)^2}{n^3}(\lambda + \frac{2n}{n-1}\lambda^2)$, we complete the proof. Finally, note that the formula for $\mathbb{V}\text{ar}(s^2)$ from the Poisson distribution slightly deviates from the one given by Mattner (1996, p. 1270). □

Theorem 5.17 (Efficiency of the sandwich estimator for a true exponential model). *Consider a sample of n independently identically $Expo(\lambda)$ distributed random variables $y_1, \ldots, y_n$ with parameter λ. The model-based variance of $\hat{\lambda}$ is estimated by $\hat{A}(\hat{\lambda}) = \bar{y}^2/n$, and the robust variance estimator is given by $\hat{C}(\hat{\lambda}) = s^2 = \frac{1}{n}\sum_{i=1}^{n}(y_i - \bar{y})^2$. If the exponential model is true, the relative efficiency of the sandwich estimator, i.e., the ratio of the variances of the model-based and the robust variance estimator, is*

$$\frac{\mathbb{V}ar(\hat{A})}{\mathbb{V}ar(\hat{C})} = \frac{n^2}{(n-1)^2} \cdot \frac{2\lambda^2}{9 - \frac{n-3}{n-1}},$$

and the asymptotic relative efficiency is

$$\lim_{n\to\infty} \frac{\mathbb{V}ar(\hat{A})}{\mathbb{V}ar(\hat{C})} = \frac{\lambda^2}{4}.$$

Proof. Mean and variance of the exponential distribution are $1/\lambda$ and $1/\lambda^2$, respectively. The ML estimator of the exponential model is $\hat{\lambda} = 1/\bar{y}$, and with $\hat{\boldsymbol{D}} = \mathbf{1}_n$, $\hat{\boldsymbol{\Sigma}}^{-1} = \operatorname{diag}(\hat{\lambda}^2)$, and $\hat{\boldsymbol{\Omega}} = \operatorname{diag}(\hat{\varepsilon}_i^2) = \operatorname{diag}\big((y_i - 1/\hat{\lambda})^2\big)$, one obtains $\hat{A}(\hat{\lambda}) = 1/\hat{\lambda}^2$, $\hat{B}(\hat{\lambda}) = \frac{1}{n}\sum_{i=1}^n \big((y_i - 1/\hat{\lambda})^2/\hat{\lambda}^4\big)$. The robust variance estimator of $\hat{\lambda}^2$ based on the assumed exponential distribution. Therefore, the robust covariance is given by

$$\begin{aligned}\hat{C}(\hat{\lambda}) &= \hat{A}^{-1}(\hat{\lambda})\,\hat{B}(\hat{\lambda})\,\hat{A}^{-1}(\hat{\lambda}) = \hat{\lambda}^2\left(\tfrac{1}{n}\sum\nolimits_{i=1}^n \tfrac{(y_i - 1/\hat{\lambda})^2}{\hat{\lambda}^4}\right)\hat{\lambda}^2 \\ &= \tfrac{1}{n}\sum\nolimits_{i=1}^n (y_i - 1/\hat{\lambda})^2 = \tfrac{1}{n}\sum\nolimits_{i=1}^n (y_i - \bar{y})^2\,.\end{aligned}$$

With $\mathbb{V}\mathrm{ar}(s^2)$ from Eq. 5.15, $m_2 = 2/\lambda^2$, $\mu_2 = 1/\lambda^2$, and $\mu_4 = 9/\lambda^4$, we obtain $\mathbb{V}\mathrm{ar}(\hat{A}) = \mathbb{V}\mathrm{ar}(\bar{y}^2) = m_2/n = 2/(n\lambda^2)$ and

$$\mathbb{V}\mathrm{ar}\big(\hat{C}(\hat{\lambda})\big) = \mathbb{V}\mathrm{ar}(s^2) = \tfrac{(n-1)^2}{n^3}\left(\tfrac{9}{\lambda^4} - \tfrac{n-3}{n-1}\tfrac{1}{\lambda^4}\right).$$

This completes the proof. □

Next, we consider the classical linear model as discussed by Carroll et al. (1998) and Kauermann and Carroll (2001). Assume that the classical homoscedastic multiple linear regression model

$$y_i = \boldsymbol{x}_i'\boldsymbol{\beta} + \varepsilon_i\,, \quad \text{with} \quad \varepsilon_i \sim N(0, \sigma^2)$$

with non-stochastic covariates $\boldsymbol{x}_i \in \mathbb{R}^p$ and $\boldsymbol{\beta} \in \mathbb{R}^p$ for $i = 1, \ldots, n$ is the true model, and without loss of generality assume that σ^2 is known. The assumed model is the heteroscedastic regression model

$$y_i = \boldsymbol{x}_i'\boldsymbol{\beta} + \varepsilon_i\,, \quad \text{with} \quad \varepsilon_i \sim N(0, \sigma_i^2)\,.$$

We are interested in the relative efficiency of the Mancl and DeRouen (2001) bias corrected robust variance estimator compared with the model-based variance estimator for linear combinations $\boldsymbol{c}'\boldsymbol{\beta}$ of the parameter vector $\boldsymbol{\beta}$ for $\boldsymbol{c} \in \mathbb{R}^p$. In the bias correction, $\hat{\tilde{\varepsilon}}_i = \hat{\varepsilon}_i/(1 - h_{ii})^{1/2}$ replaces $\hat{\varepsilon}_i$ in the OPG of the robust variance estimator (see Remark 5.5). Finally, let $a_i = \boldsymbol{c}'(\boldsymbol{X}'\boldsymbol{X})^{-1}\boldsymbol{x}_i$.

Theorem 5.18 (Linear regression with heteroscedastic variance). *The relative efficiency of the model-based variance estimator of the linear combination $\boldsymbol{c}'\hat{\boldsymbol{\beta}}$ compared with the corresponding bias corrected robust variance estimator is given by*

$$\frac{\mathbb{V}ar\Big(\hat{A}\big(\boldsymbol{c}'\hat{\boldsymbol{\beta}}\big)\Big)}{\mathbb{V}ar\Big(\hat{C}_{MD}\big(\boldsymbol{c}'\hat{\boldsymbol{\beta}}\big)\Big)} = \left(\tfrac{1}{n}\sum\nolimits_{i=1}^n a_i^2\right)^2\Big/\left(\tfrac{1}{n}\left(\sum\nolimits_{i=1}^n a_i^4 + \sum\nolimits_{i\neq j} a_i^2 a_j^2 \tilde{h}_{ij}\right)\right). \quad (5.16)$$

Remark 5.19. Equation 5.16 can be simplified by using an analogous argument as Mancl and DeRouen (2001) in their approximation for obtaining the bias correction (Remark 5.5). First, we note that $\boldsymbol{H}$ is an orthogonal projection matrix of rank p, thus $\mathrm{tr}(\boldsymbol{H}) = p$. Ideally, all observations have the same leverage, and the off-diagonal elements of $\boldsymbol{H}$ might therefore be negligible. In this case, the last term can be omitted, and Eq. (5.16) reduces to

$$\frac{\mathbb{V}\mathrm{ar}\Big(\hat{A}\big(\boldsymbol{c}'\hat{\boldsymbol{\beta}}\big)\Big)}{\mathbb{V}\mathrm{ar}\Big(\hat{C}_{MD}\big(\boldsymbol{c}'\hat{\boldsymbol{\beta}}\big)\Big)} \approx \Big(\tfrac{1}{n}\sum\nolimits_{i=1}^{n} a_i^2\Big)^2 \Big/ \Big(\tfrac{1}{n}\sum\nolimits_{i=1}^{n} a_i^4\Big) . \tag{5.17}$$

In any case, the variability of the robust variance estimator is greater than the variability of the model based variance estimator, and this finding can be generalized to the generalized linear model (Kauermann and Carroll, 2001).

Proof. The model-based and robust variance estimators of $\hat{\boldsymbol{\beta}}$ have been given in Example 5.12. By using the bias correction of Mancl and DeRouen (2001, see Remark 5.5), we obtain the bias corrected robust variance estimator $\hat{\boldsymbol{C}}_{MD}(\hat{\boldsymbol{\beta}}) = (\boldsymbol{X}'\boldsymbol{X})^{-1}\big(\sum_{i=1}^{n} \boldsymbol{x}_i\boldsymbol{x}_i'\hat{\hat{\varepsilon}}_i^2\big)(\boldsymbol{X}'\boldsymbol{X})^{-1}$ with $\hat{\hat{\varepsilon}}_i = \hat{\varepsilon}_i/(1-h_{ii})^{1/2}$.

Using the normality assumption, we obtain

$$\mathbb{V}\mathrm{ar}\big(\hat{A}(\boldsymbol{c}'\hat{\boldsymbol{\beta}})\big) = 2\sigma^4\big(\boldsymbol{c}'(\boldsymbol{X}'\boldsymbol{X})^{-1}\boldsymbol{c}\big)^2/n = 2\sigma^4\Big(\sum_{i=1}^{n} a_i^2\Big)^2/n\,, \tag{5.18}$$

with $a_i = \boldsymbol{c}'(\boldsymbol{X}'\boldsymbol{X})^{-1}\boldsymbol{x}_i$. Because $\mathbb{V}\mathrm{ar}\big(\hat{\hat{\varepsilon}}_i^2\big) = 2\sigma^4$ and $\mathbb{C}\mathrm{ov}\big(\hat{\hat{\varepsilon}}_i, \hat{\hat{\varepsilon}}_j\big) = 2\tilde{h}_{ij}\,\sigma^4$ for $i \neq j$ and $\tilde{h}_{ij} = h_{ij}/\sqrt{(1-h_{ii})(1-h_{jj})}$, we similarly derive

$$\begin{aligned}\mathbb{V}\mathrm{ar}\big(\hat{C}_{MD}(\boldsymbol{c}'\hat{\boldsymbol{\beta}})\big) &= \sum_{i=1}^{n} a_i^4 \mathbb{V}\mathrm{ar}\big(\hat{\hat{\varepsilon}}_i^2\big) + \sum_{i\neq j} a_i^2 a_j^2 \mathbb{C}\mathrm{ov}\big(\hat{\hat{\varepsilon}}_i, \hat{\hat{\varepsilon}}_j\big) \\ &= 2\sigma^4\Big(\sum_{i=1}^{n} a_i^4 + \sum_{i\neq j} a_i^2 a_j^2 \tilde{h}_{ij}\Big)\,,\end{aligned}$$

which completes the proof. □

Theorem 5.18 is illustrated in a simple example using a parallel group controlled clinical trial with 1:1 randomization.

Example 5.20. Consider a parallel group controlled clinical trial where half of the patients are randomized to a new treatment $x_T = 1$, and the other half receive the standard treatment $x_S = 0$. For simplicity, we assume that n is even. The $n \times 2$ design matrix $\boldsymbol{X}$ is given by

$$\boldsymbol{X} = \begin{pmatrix} 1 \ldots 1\; 1 \ldots 1 \\ 0 \ldots 0\; 1 \ldots 1 \end{pmatrix}' .$$

We are interested in testing the treatment effect so that $\boldsymbol{c} = (0\ 1)'$. With

$$(\boldsymbol{X}'\boldsymbol{X})^{-1} = \begin{pmatrix} \frac{2}{n} & -\frac{2}{n} \\ -\frac{2}{n} & \frac{4}{n} \end{pmatrix} \quad \text{and} \quad \begin{aligned} a_S &= \boldsymbol{c}'(\boldsymbol{X}'\boldsymbol{X})^{-1}(1\ x_S)' = -\tfrac{2}{n} \\ a_T &= \boldsymbol{c}'(\boldsymbol{X}'\boldsymbol{X})^{-1}(1\ x_T)' = \tfrac{2}{n} \end{aligned}$$

we obtain $\sum_{i=1}^n a_i^2 = 4/n$ and $\sum_{i=1}^n a_i^4 = 16/n^3$. The hat matrix $\boldsymbol{H} = \boldsymbol{X}(\boldsymbol{X}'\boldsymbol{X})^{-1}\boldsymbol{X}'$ is a block diagonal matrix with block length $n/2$ and block entries $2/n$. Subsequently, $\tilde{\boldsymbol{H}}$, the matrix consisting of elements $\tilde{h}_{ij}$, is a block diagonal matrix with the same block length and block entries $1/(n/2-1)$, and $\sum_{i \neq j} a_i^2 a_j^2 \tilde{h}_{ij} = 2 \cdot \frac{n}{2}\big(\frac{n}{2}-1\big)\frac{4}{n^2} \cdot \frac{4}{n^2}\frac{1}{n/2-1} = \frac{16}{n^3}$.

Because $\big(\frac{1}{n}\sum_{i=1}^n a_i^2\big)^2 = 16/n^4$ and $\frac{1}{n}\big(\sum_{i=1}^n a_i^4 + \sum_{i \neq j} a_i^2 a_j^2 \tilde{h}_{ij}\big) = \frac{1}{n}\big(\frac{16}{n^3} + \frac{16}{n^3}\big) = \frac{32}{n^4}$, the relative efficiency of the robust variance estimator compared with the model-based variance estimator is only $1/2 = 50\%$.

Note that the approximation to the relative efficiency from Remark 5.19 gives a relative efficiency of 100% because the term $\tilde{h}_{ij}$ is neglected.

5.4.2 Bias corrections and small sample adjustments

The small sample properties of the robust variance estimator were investigated in a series of papers around 1990 in Monte-Carlo simulations (see, e.g., Emrich and Piedmonte, 1992; Gunsolley et al., 1995; Paik, 1988; Sharples and Breslow, 1992). Indeed, the robust variance estimator tends to underestimate the variance of the regression coefficients, especially in the case of small samples (Mancl and DeRouen, 2001). Basically, four different approaches have been considered for improving the small sample performance of the robust variance estimator. First, jack-knife estimators (Lipsitz et al., 1994a, see Theorem 5.21) and bootstrap estimators (Lancaster, 2003) are alternatives to the robust variance estimator, and they keep the nominal test level of 5% well if the sample size is not smaller than 20 (Mancl and DeRouen, 2001). Second, in the previous section, we have seen that the sandwich estimator has increased variability compared with the model-based variance estimator (see Theorem 5.18 and Remark 5.19), and one may account for this extra variability (see, e.g., Kauermann and Carroll, 2001; Pan and Wall, 2002; for a brief overview, see, e.g., Dahmen and Ziegler, 2004). Third, for statistical testing and the construction of confidence intervals, the use of the t distribution or the F distribution has been proposed (Fay et al., 1998; Pan and Wall, 2002). Finally, as already seen above (Theorem 5.4), the robust variance estimator is biased, and the small sample properties of the sandwich estimator can be improved by bias corrections.

Theorem 5.21 (Bias correction of the robust covariance matrix). *Using the notations and assumptions from above, the following estimators of the robust covariance matrix are less biased than the estimator of the robust covariance matrix from Theorem 5.2, 4.:*

1. *The bias corrected robust covariance matrix estimator according to Mancl and DeRouen (2001) is given by*

$$\hat{\boldsymbol{C}}_{MD}(\boldsymbol{\beta}) = \hat{\boldsymbol{A}}(\boldsymbol{\beta})^{-1}\hat{\boldsymbol{B}}_{MD}(\boldsymbol{\beta})\hat{\boldsymbol{A}}(\boldsymbol{\beta})^{-1}\,,$$

where $\hat{\boldsymbol{B}}_{MD}(\boldsymbol{\beta})$ *is given by*

$$\hat{\boldsymbol{B}}_{MD}(\boldsymbol{\beta}) = \frac{1}{n}\sum_{i=1}^{n}\hat{\boldsymbol{D}}_i'\boldsymbol{\Sigma}_i^{-1}(\boldsymbol{I}-\hat{\boldsymbol{H}}_{ii})^{-1}\hat{\boldsymbol{\Omega}}_i(\boldsymbol{I}-\hat{\boldsymbol{H}}_{ii})^{-1}\boldsymbol{\Sigma}_i^{-1}\hat{\boldsymbol{D}}_i\,,$$

and $\hat{\boldsymbol{H}}_{ii} = \hat{\boldsymbol{D}}_i\hat{\boldsymbol{A}}^{-1}\hat{\boldsymbol{D}}_i'\boldsymbol{\Sigma}_i^{-1}$.

2. *The modified Fay and Graubard bias corrected robust covariance matrix estimator is given by*

$$\hat{\boldsymbol{C}}_{mFG}(\boldsymbol{\beta}) = \hat{\boldsymbol{A}}(\boldsymbol{\beta})^{-1}\hat{\boldsymbol{B}}_{mFG}(\boldsymbol{\beta})\hat{\boldsymbol{A}}(\boldsymbol{\beta})^{-1}\,,$$

where $\hat{\boldsymbol{B}}_{mFG}(\boldsymbol{\beta})$ *is given by*

$$\boldsymbol{B}_{mFG}(\boldsymbol{\beta}) = \frac{1}{n}\sum_{i=1}^{n}\tilde{\boldsymbol{H}}_i\hat{\boldsymbol{D}}_i'\boldsymbol{\Sigma}_i^{-1}\hat{\boldsymbol{\Omega}}_i\boldsymbol{\Sigma}_i^{-1}\hat{\boldsymbol{D}}_i\tilde{\boldsymbol{H}}_i'\,,$$

and $\tilde{\boldsymbol{H}}_i = (\boldsymbol{I}-\hat{\boldsymbol{A}}_i\hat{\boldsymbol{A}}^{-1})^{-1/2}$.

Remark 5.22. The modified Fay and Graubard version of the bias correction can be used if $\boldsymbol{I}-\boldsymbol{A}_i\boldsymbol{A}^{-1}$ is positive definite, and it differs from the original version of Fay and Graubard (2001). First, these authors specifically used the Hessian matrices $\boldsymbol{W}_i$ and $\boldsymbol{W}$ in the approximation instead of the Fisher information matrices $\boldsymbol{A}_i$ and $\boldsymbol{A}$, and they pointed out that their equivalent of $\tilde{\boldsymbol{H}}_i$ is not necessarily symmetric (Fay and Graubard, 2001, p. 1199). They therefore suggested the replacement of $\boldsymbol{H}_i$ by a diagonal matrix with the jjth element being estimated by $(1-\min(b,[\hat{\boldsymbol{A}}_i\hat{\boldsymbol{A}}^{-1}]_{jj}))^{-1/2}$. This bound b is a practical necessity to prevent from extreme adjustments when $[\hat{\boldsymbol{A}}_i\hat{\boldsymbol{A}}^{-1}]_{jj}$ is close to 1, and Fay and Graubard (2001) arbitrarily used $b = 0.75$ in their Monte-Carlo simulations. Interestingly, when they considered GEE as a special case, they replaced the Hessian matrices with the Fisher information matrices (Fay and Graubard, 2001, p. 1200).

———

For other modifications to the robust variance estimator the reader may refer to the literature (see, e.g., Lu et al., 2007; Morel et al., 2003; Pan, 2001; Wang and Long, 2011); for a review, see Dahmen and Ziegler (2004).

Proof.
1.: Mancl and DeRouen (2001) assumed that the last term of Eq. 5.5 is negligible. With the arguments of the last bullet point in Remark 5.5, the

proof is completed.
2.: This bias correction is based on the first part of Theorem 5.4. Specifically, Fay and Graubard (2001) argued that the elements of the second term of Eq. 5.4 could well be different from $\mathbf{0}$ so that this term should not be neglected. To derive a tractable expression, Fay and Graubard (2001) assumed that the Fisher information matrix $\boldsymbol{A}_i$ is within a scale factor of $\boldsymbol{\Omega}_i$, i.e., $\boldsymbol{A}_i \approx c\boldsymbol{\Omega}_i$ for all i and some constant c. As pointed out by Dahmen and Ziegler (2004), this proportionality assumption is unrealistic for misspecified variance structures. However, it leads to a substantial simplification, and Eq. 5.4 may be rewritten as

$$\mathbb{E}^{\boldsymbol{X}}\mathbb{E}^{\boldsymbol{y}}(\hat{\boldsymbol{B}}_i) = \mathbb{E}^{\boldsymbol{X}}\mathbb{E}^{\boldsymbol{y}}(\hat{\boldsymbol{u}}_i\hat{\boldsymbol{u}}_i') \approx \left(\boldsymbol{I} - \boldsymbol{A}_i\boldsymbol{A}^{-1}\right)\boldsymbol{B}_i$$

because the product $\boldsymbol{A}^{-1}\boldsymbol{B}$ cancels out. Both $\boldsymbol{A}_i$ and $\boldsymbol{A}$ are symmetric so that $\hat{\boldsymbol{B}}_i$ can be bias corrected by $\hat{\boldsymbol{B}}_{FG,i} = \tilde{\boldsymbol{H}}_i\hat{\boldsymbol{B}}_i\tilde{\boldsymbol{H}}_i'$ with $\tilde{\boldsymbol{H}}_i = (\boldsymbol{I} - \hat{\boldsymbol{A}}_i\hat{\boldsymbol{A}}^{-1})^{-1/2}$. This completes the proof. □

Theorem 5.23 (Asymptotic equivalence of the one-step jack-knife estimator of covariance and the robust covariance estimator). *Let $\hat{\boldsymbol{\beta}}$ be the PML estimator, and let $\hat{\boldsymbol{\beta}}(i)$ be the PML estimator after deleting cluster i and performing one step of the modified Fisher scoring step. The update step is given by*

$$\hat{\boldsymbol{\beta}}^{(i)} = \hat{\boldsymbol{\beta}} + \left(\sum_{i=1}^{n} \hat{\boldsymbol{D}}_{(i)}'\boldsymbol{\Sigma}_{(i)}^{-1}\hat{\boldsymbol{D}}_{(i)}\right)^{-1}\left(\sum_{i=1}^{n} \hat{\boldsymbol{D}}_{(i)}'\boldsymbol{\Sigma}_{(i)}^{-1}\hat{\varepsilon}_{(i)}\right). \tag{5.19}$$

The one-step jack-knife estimator of covariance

$$\frac{n-p}{n}\left(\hat{\boldsymbol{\beta}}_{(i)} - \hat{\boldsymbol{\beta}}\right)\left(\hat{\boldsymbol{\beta}}_{(i)} - \hat{\boldsymbol{\beta}}\right)'$$

is asymptotically equivalent to $\hat{\boldsymbol{C}}$, the estimator of the robust covariance matrix.

Proof. First, we note that $\sum_{i=1}^{n} \hat{\boldsymbol{D}}_{(i)}'\boldsymbol{\Sigma}_{(i)}^{-1}\hat{\boldsymbol{D}}_{(i)} = -\hat{\boldsymbol{D}}_i'\boldsymbol{\Sigma}_i^{-1}\hat{\boldsymbol{D}}_i$ because $\hat{\boldsymbol{u}}(\hat{\boldsymbol{\beta}}) = \mathbf{0}$ at $\hat{\boldsymbol{\beta}}$. Therefore, Eq. 5.19 can be written as

$$\hat{\boldsymbol{\beta}}^{(i)} - \hat{\boldsymbol{\beta}} = -\left(\sum_{i=1}^{n} \hat{\boldsymbol{D}}_{(i)}'\boldsymbol{\Sigma}_{(i)}^{-1}\hat{\boldsymbol{D}}_{(i)}\right)^{-1}\hat{\boldsymbol{D}}_i'\boldsymbol{\Sigma}_i^{-1}\hat{\varepsilon}_i\,, \tag{5.20}$$

and we obtain

$$\left(\hat{\boldsymbol{\beta}}^{(i)} - \hat{\boldsymbol{\beta}}\right)\left(\hat{\boldsymbol{\beta}}^{(i)} - \hat{\boldsymbol{\beta}}\right)' = \sum_{i=1}^{n} \left(\left(\sum_{i=1}^{n} \hat{\boldsymbol{D}}'_{(i)} \boldsymbol{\Sigma}_{(i)}^{-1} \hat{\boldsymbol{D}}_{(i)} \right)^{-1} \left(\hat{\boldsymbol{D}}'_i \boldsymbol{\Sigma}_i^{-1} \hat{\boldsymbol{\varepsilon}}_i \hat{\boldsymbol{\varepsilon}}'_i \boldsymbol{\Sigma}_i^{-1} \hat{\boldsymbol{D}}'_i \right) \left(\sum_{i=1}^{n} \hat{\boldsymbol{D}}'_{(i)} \boldsymbol{\Sigma}_{(i)}^{-1} \hat{\boldsymbol{D}}_{(i)} \right)^{-1} \right).$$

By noting that $\sum_{i=1}^{n} \hat{\boldsymbol{D}}'_{(i)} \boldsymbol{\Sigma}_{(i)}^{-1} \hat{\boldsymbol{D}}_{(i)}$ is asymptotically equivalent to $\hat{\boldsymbol{D}}' \boldsymbol{\Sigma}^{-1} \hat{\boldsymbol{D}}$, the proof is complete. □

Remark 5.24.

- An alternative approach for updating parameter estimates is the modified Iteratively (Re–) Weighted Least Squares (IWLS) algorithm, which is equivalent to the modified Fisher scoring. Indeed, pre-multiplication of $\hat{\boldsymbol{\beta}}$ by $\left(\sum_{i=1}^{n} \hat{\boldsymbol{D}}'_{(i)} \boldsymbol{\Sigma}_{(i)}^{-1} \hat{\boldsymbol{D}}_{(i)}\right)^{-1} \left(\sum_{i=1}^{n} \hat{\boldsymbol{D}}'_{(i)} \boldsymbol{\Sigma}_{(i)}^{-1} \hat{\boldsymbol{D}}_{(i)}\right)$ yields the update formula of the modified IWLS algorithm:

$$\hat{\boldsymbol{\beta}}^{(i)} = \left(\sum_{i=1}^{n} \hat{\boldsymbol{D}}'_{(i)} \boldsymbol{\Sigma}_{(i)}^{-1} \hat{\boldsymbol{D}}_{(i)} \right)^{-1} \left(\sum_{i=1}^{n} \hat{\boldsymbol{D}}'_{(i)} \boldsymbol{\Sigma}_{(i)}^{-1} \left(\hat{\boldsymbol{D}}_{(i)} \hat{\boldsymbol{\beta}} + \hat{\boldsymbol{\varepsilon}}_{(i)} \right) \right).$$

- A disadvantage of Eq. (5.19) is that it involves the calculation of deletion statistics, e.g., $\hat{\boldsymbol{D}}_{(i)}$ which are typically computed using all available data. An alternative representation of $\left(\hat{\boldsymbol{\beta}}^{(i)} - \hat{\boldsymbol{\beta}}\right)$ not involving deletion statistics is

$$\hat{\boldsymbol{A}}^{-1} \hat{\boldsymbol{Z}}'_i \hat{\boldsymbol{K}}_i \boldsymbol{\Sigma}_i^{-1/2} \hat{\boldsymbol{\varepsilon}}_i \quad \text{with} \quad \boldsymbol{K}_i = \boldsymbol{I} + \left(\boldsymbol{I} - \hat{\boldsymbol{Z}}_i \hat{\boldsymbol{A}}^{-1} \hat{\boldsymbol{Z}}'_i\right)^{-1} \hat{\boldsymbol{Z}}_i \hat{\boldsymbol{A}}^{-1} \hat{\boldsymbol{Z}}'_i,$$

with $\hat{\boldsymbol{Z}}_i = \boldsymbol{\Sigma}_i^{-1/2} \hat{\boldsymbol{D}}_i$, where $\boldsymbol{\Sigma}_i^{1/2}$ is a root of $\boldsymbol{\Sigma}_i$ as in the proof of Theorem 5.2.

Proof (Second bullet point of Remark 5.24). Using the notation from above, Eq. 5.20 can be written as

$$\hat{\boldsymbol{\beta}}^{(i)} - \hat{\boldsymbol{\beta}} = -\hat{\boldsymbol{A}}_{(i)}^{-1} \hat{\boldsymbol{Z}}'_i \boldsymbol{\Sigma}_i^{-1/2} \hat{\boldsymbol{\varepsilon}}_i .$$

Application of the update formula for symmetric matrices (Cook and Weisberg, 1982, Appendix, Eq. A.2.1) yields

$$\hat{\boldsymbol{A}}_{(i)}^{-1} = (\hat{\boldsymbol{Z}}'_{(i)} \hat{\boldsymbol{Z}}_{(i)})^{-1} = (\hat{\boldsymbol{Z}}'\hat{\boldsymbol{Z}})^{-1} + (\hat{\boldsymbol{Z}}'\hat{\boldsymbol{Z}})^{-1} \hat{\boldsymbol{Z}}'_i (\boldsymbol{I} - \hat{\boldsymbol{Z}}_i (\hat{\boldsymbol{Z}}'\hat{\boldsymbol{Z}})^{-1} \hat{\boldsymbol{Z}}'_i)^{-1} \hat{\boldsymbol{Z}}_i (\hat{\boldsymbol{Z}}'\hat{\boldsymbol{Z}})^{-1} .$$

Appropriate matrix multiplications complete the proof. □

Chapter 6
Quasi generalized pseudo maximum likelihood method based on the linear exponential family

PML1 allows estimation of the correctly specified mean structure either when the assumed distribution includes no nuisance parameter or when an additional nuisance parameter, such as the covariance matrix from the normal distribution, is fixed. For applications, this assumption might be unrealistic because the a priori guess for the nuisance parameter is rarely good. Furthermore, the more similar the chosen nuisance parameter is to the optimal nuisance parameter, the more efficient the estimator will be. This leads to the idea that the user specifies a specific structure for the nuisance parameter $\boldsymbol{\Psi}$. The nuisance parameter $\hat{\boldsymbol{\Psi}}$ is estimated in the first step, and the parameter of interest $\boldsymbol{\beta}$ is estimated in the second given the estimated nuisance parameter $\hat{\boldsymbol{\Psi}}$. Again, we stress that the parameter $\boldsymbol{\Psi}$, typically determining the covariance of correlated observations or an additional variance parameter, such as in overdispersed models, is considered to be nuisance. Models that aim at estimating both the mean and the association parameters as the parameter of interest are discussed in the next two chapters. The extension of PML1 estimation considered in this chapter allows for nuisance parameters; it is a quasi generalization of the PML1 approach and therefore termed the quasi generalized pseudo maximum likelihood (QGPML) method.

An important aspect of this approach is how a consistent estimator of the parameter vector $\boldsymbol{\beta}$ of the mean structure and its variance can be obtained although additional variability is introduced through the estimation of a possibly misspecified nuisance parameter $\boldsymbol{\Psi}$. The surprising result is that under suitable regularity conditions, one need not account for the extra variability introduced by estimating the nuisance parameter. The QGPML approach can be considered an extension of the work by Burguete et al. (1982), and it has been discussed in detail by Gourieroux et al. (1984b). However, Gourieroux and colleagues (Gourieroux et al., 1984b; Gourieroux and Monfort, 1993, 1995a) did not discuss a possible misspecification of the nuisance parameter in their work although they already formulated more general results in a different context.

This chapter is organized as follows. We first define the QGPML estimator, and second derive its asymptotic properties (Sect. 6.2). We next illustrate the QGPML approach in a series of examples (Sect. 6.3). Specifically, we consider generalized estimating equations of order 1 (GEE1) with estimated working covariance matrix and estimated working correlation matrix. Finally, we briefly discuss extensions to time dependent parameters and ordinal response variables (Sect. 6.4).

6.1 Definition

Consider a sample of n independently distributed T-dimensional random vectors $\boldsymbol{y}_i$, $i = 1, \ldots, n$, and $\boldsymbol{X}_i$ is the $T \times p$ matrix of stochastic and/or fixed explanatory variables of subject i. The true but unknown density (or probability mass function for discrete random vectors) of $\boldsymbol{y}_i$ given $\boldsymbol{X}_i$ is denoted by $f^\star(\boldsymbol{y}_i|\boldsymbol{X}_i)$ with conditional expectation $\boldsymbol{\mu}_i = \mathbb{E}_{f^\star}(\boldsymbol{y}_i|\boldsymbol{X}_i||\boldsymbol{\beta}_0) = \boldsymbol{g}(\boldsymbol{X}_i\boldsymbol{\beta})$ and conditional covariance matrix $\mathbb{V}\mathrm{ar}_{f^\star}(\boldsymbol{y}_i|\boldsymbol{X}_i) = \boldsymbol{\Omega}^\star(\boldsymbol{X}_i)$.

The true density $f^\star$ may differ from the assumed density f. The assumed conditional pseudo density f of $\boldsymbol{y}_i$ given $\boldsymbol{X}_i$ is parameterized in the $p \times 1$ parameter vector $\boldsymbol{\beta}$ so that $\mathbb{E}_f(\boldsymbol{y}_i|\boldsymbol{X}_i||\boldsymbol{\beta})$. The mean is assumed to be correctly specified, i.e.,

$$\mathbb{E}_{f^\star}(\boldsymbol{y}_i|\boldsymbol{X}_i) = \mathbb{E}_f(\boldsymbol{y}_i|\boldsymbol{X}_i||\boldsymbol{\beta}_0)\,.$$

Estimation is based on a pseudo density from the linear exponential family with nuisance parameter $\boldsymbol{\Psi}$, and the assumed density of cluster i is

$$f(\boldsymbol{y}_i|\boldsymbol{X}_i||\boldsymbol{\mu}_i, \boldsymbol{\Psi}_i) = \exp\big(a(\boldsymbol{\mu}_i, \boldsymbol{\Psi}_i) + b(\boldsymbol{y}_i, \boldsymbol{\Psi}_i) + \boldsymbol{c}(\boldsymbol{\mu}_i, \boldsymbol{\Psi}_i)'\boldsymbol{y}_i\big)\,,$$

with a possibly cluster specific nuisance parameter $\boldsymbol{\Psi}_i$. According to Property 1.3, the inverse variance matrix of cluster i is given by $\boldsymbol{\Sigma}_i^{-1} = \partial \boldsymbol{c}(\boldsymbol{\mu}_i, \boldsymbol{\Psi}_i)'/\partial\boldsymbol{\mu}$ for the assumed density. For many univariate and multivariate distributions belonging to the linear exponential family, the nuisance parameter $\boldsymbol{\Psi}_i$ can be written as a differentiable function $\boldsymbol{G}$ of $\boldsymbol{\mu}_i$ and $\boldsymbol{\Sigma}_i$ according to the implicit function theorem $\boldsymbol{\Psi}_i = \boldsymbol{G}(\boldsymbol{\mu}_i, \boldsymbol{\Sigma}_i)$. In most applications, $\boldsymbol{\Sigma}_i$ is chosen as a function of the explanatory variables $\boldsymbol{X}_i$, the parameter vector of interest $\boldsymbol{\beta}$, and an additional $q \times 1$ parameter vector $\boldsymbol{\alpha}$, so that $\boldsymbol{\Psi}_i$ may also be written as $\boldsymbol{\Psi}_i = \boldsymbol{G}\big(\boldsymbol{\mu}(\boldsymbol{X}_i, \boldsymbol{\beta}_0), \boldsymbol{\Sigma}(\boldsymbol{X}_i, \boldsymbol{\beta}_0, \boldsymbol{\alpha}_0)\big)$.

Estimation may now proceed in two steps: In the first step, estimates $\tilde{\boldsymbol{\beta}}$ and $\tilde{\boldsymbol{\alpha}}$ of $\boldsymbol{\beta}_0$ and $\boldsymbol{\alpha}_0$ are obtained. In practice, $\tilde{\boldsymbol{\beta}}$ and $\tilde{\boldsymbol{\alpha}}$ are often obtained by simple method of moments or least squares estimators yielding consistent but possibly inefficient estimates. Based on estimates $\tilde{\boldsymbol{\alpha}}$ and $\tilde{\boldsymbol{\beta}}$, the nuisance parameter $\boldsymbol{\Psi}_i$ is fixed:

$$\tilde{\boldsymbol{\Psi}}_i = \boldsymbol{G}\big(\boldsymbol{\mu}(\boldsymbol{X}_i, \tilde{\boldsymbol{\beta}}), \boldsymbol{\Sigma}(\boldsymbol{X}_i, \tilde{\boldsymbol{\beta}}, \tilde{\boldsymbol{\alpha}})\big)\,.$$

In the second step, the QGPML estimator $\hat{\boldsymbol{\beta}}$ for $\boldsymbol{\beta}_0$ is computed using the estimate $\tilde{\boldsymbol{\Psi}}_i$ for $\boldsymbol{\Psi}_i$. This two-step procedure is a generalization of the method proposed by Cochrane and Orcutt (1949) to compute the least squares estimator in an autoregressive linear model.

In summary, the stochastic model of $\boldsymbol{y}_i$ given $\boldsymbol{X}_i$ under the true model $f^\star(\boldsymbol{y}_i|\boldsymbol{X}_i||\boldsymbol{\beta}_0)$ is given by

$$\begin{aligned} \boldsymbol{y}_i &= \boldsymbol{\mu}(\boldsymbol{X}_i, \boldsymbol{\beta}_0) + \boldsymbol{\varepsilon}_i^\star \qquad \text{with} \qquad \mathbb{E}(\boldsymbol{\varepsilon}_i^\star|\boldsymbol{X}_i) = \mathbf{0}\,, \\ \mathbb{E}_{f^\star}(\boldsymbol{y}_i|\boldsymbol{X}_i) &= \mathbb{E}_f(\boldsymbol{y}_i|\boldsymbol{X}_i) = \boldsymbol{\mu}(\boldsymbol{X}_i, \boldsymbol{\beta}_0) = \boldsymbol{\mu}_i\,, \\ \mathbb{V}\mathrm{ar}_{f^\star}(\boldsymbol{y}_i|\boldsymbol{X}_i) &= \mathbb{V}\mathrm{ar}(\boldsymbol{\varepsilon}_i^\star|\boldsymbol{X}_i) = \boldsymbol{\Omega}(\boldsymbol{X}_i)\,. \end{aligned}$$

The stochastic model of $\boldsymbol{y}_i$ given $\boldsymbol{X}_i$ under the assumed model $f(\boldsymbol{y}_i|\boldsymbol{X}_i||\boldsymbol{\beta}_0)$ is given by

$$\begin{aligned} \boldsymbol{y}_i &= \boldsymbol{\mu}(\boldsymbol{X}_i, \boldsymbol{\beta}_0) + \boldsymbol{\varepsilon}_i \qquad \text{with} \qquad \mathbb{E}(\boldsymbol{\varepsilon}_i|\boldsymbol{X}_i) = \mathbf{0}\,, \\ \mathbb{E}_{f^\star}(\boldsymbol{y}_i|\boldsymbol{X}_i) &= \mathbb{E}_f(\boldsymbol{y}_i|\boldsymbol{X}_i) = \boldsymbol{\mu}(\boldsymbol{X}_i, \boldsymbol{\beta}_0) = \boldsymbol{\mu}_i\,, \\ \mathbb{V}\mathrm{ar}_f(\boldsymbol{y}_i|\boldsymbol{X}_i) &= \mathbb{V}\mathrm{ar}(\boldsymbol{\varepsilon}_i|\boldsymbol{X}_i) = \tilde{\boldsymbol{\Sigma}}(\boldsymbol{X}_i, \tilde{\boldsymbol{\beta}}, \tilde{\boldsymbol{\Psi}}_i)\,, \end{aligned}$$

where the assumed density belongs to the linear exponential family.

Definition 6.1 (QGPML estimator). A quasi generalized pseudo maximum likelihood estimator for the mean, or, briefly, QGPML estimator of $\boldsymbol{\beta}$, is any value $\hat{\boldsymbol{\beta}}$ maximizing the kernel of the normed pseudo loglikelihood function

$$\begin{aligned} l(\boldsymbol{\beta}, \tilde{\boldsymbol{\alpha}}) &= \frac{1}{n}\sum_{i=1}^{n} \ln f(\boldsymbol{y}_i|\boldsymbol{X}_i||\boldsymbol{\beta}, \tilde{\boldsymbol{\alpha}}) = \frac{1}{n}\sum_{i=1}^{n} \ln f\big(\boldsymbol{y}_i||\boldsymbol{\mu}(\boldsymbol{X}_i, \boldsymbol{\beta}), \tilde{\boldsymbol{\Psi}}_i\big) \\ &= \frac{1}{n}\sum_{i=1}^{n} \ln f\left(\boldsymbol{y}_i||\boldsymbol{\mu}(\boldsymbol{X}_i, \boldsymbol{\beta}), \boldsymbol{G}\big(\boldsymbol{\mu}(\boldsymbol{X}_i, \tilde{\boldsymbol{\beta}}), \boldsymbol{\Sigma}(\boldsymbol{X}_i, \tilde{\boldsymbol{\beta}}, \tilde{\boldsymbol{\alpha}})\big)\right)\,. \end{aligned}$$

6.2 Asymptotic properties

The following theorem summarizes the properties of the QGPML estimator. The required regularity conditions and detailed proofs can be found in Gourieroux et al. (1984b, pp. 682, 687, 692). To repeat, a fundamental assumption is that the assumed distribution belongs to the linear exponential family.

Theorem 6.2 (Properties of QGPML estimators).

1. *There asymptotically exists a QGPML estimator $\hat{\boldsymbol{\beta}}$ for $\boldsymbol{\beta}_0$.*
2. *The QGPML estimator $\hat{\boldsymbol{\beta}}$ converges almost surely to the true parameter vector $\boldsymbol{\beta}_0$.*

3. *The score vector for* $\boldsymbol{\beta}$ *is given by*

$$\boldsymbol{u}(\boldsymbol{\beta}) = \sum_{i=1}^{n} \boldsymbol{D}_i' \tilde{\boldsymbol{\Sigma}}_i^{-1} (\boldsymbol{y}_i - \boldsymbol{\mu}_i) = \boldsymbol{D}' \tilde{\boldsymbol{\Sigma}}^{-1} (\boldsymbol{y} - \boldsymbol{\mu}) \,,$$

where $\boldsymbol{D}$ *is the stacked matrix of the* $\boldsymbol{D}_i$, $\tilde{\boldsymbol{\Sigma}}$ *is the block diagonal matrix of the* $\tilde{\boldsymbol{\Sigma}}_i$, *and* $\boldsymbol{y}$ *and* $\boldsymbol{\mu}$ *are the stacked vectors* $\boldsymbol{y}_i$ *and* $\boldsymbol{\mu}_i$, *respectively.* $\tilde{\boldsymbol{\Sigma}}_i = \tilde{\boldsymbol{\Sigma}}_i(\boldsymbol{X}_i, \tilde{\boldsymbol{\beta}}, \tilde{\boldsymbol{\Psi}}_i)$ *is the working covariance matrix fixed at* $\tilde{\boldsymbol{\beta}}$ *and* $\tilde{\boldsymbol{\Psi}}_i$.

4. *The QGPML estimator* $\hat{\boldsymbol{\beta}}$ *for* $\boldsymbol{\beta}_0$ *is asymptotically normal. More specifically,*

$$\sqrt{n}(\hat{\boldsymbol{\beta}} - \boldsymbol{\beta}_0) \overset{a}{\sim} N\Big(\boldsymbol{0}, \boldsymbol{A}(\boldsymbol{\beta}_0, \boldsymbol{\alpha}_0)^{-1} \boldsymbol{B}(\boldsymbol{\beta}_0, \boldsymbol{\alpha}_0) \boldsymbol{A}(\boldsymbol{\beta}_0, \boldsymbol{\alpha}_0)^{-1}\Big) \,, \qquad (6.1)$$

where $\boldsymbol{A}(\boldsymbol{\beta}, \boldsymbol{\alpha}) = \mathbb{E}^{\boldsymbol{X}}\big(\mathbb{E}^{\boldsymbol{y}}_{f^\star} - \boldsymbol{W}_i\big) = \mathbb{E}^{\boldsymbol{X}}(\boldsymbol{D}_i' \boldsymbol{\Sigma}_i^{-1} \boldsymbol{D}_i)$ *is the Fisher information matrix and* $\boldsymbol{B}(\boldsymbol{\beta}, \boldsymbol{\alpha}) = \mathbb{E}^{\boldsymbol{X}} \mathbb{E}^{\boldsymbol{y}}_{f^\star}(\boldsymbol{u}_i(\boldsymbol{\beta}) \boldsymbol{u}_i(\boldsymbol{\beta})') = \mathbb{E}^{\boldsymbol{X}}(\boldsymbol{D}_i' \boldsymbol{\Sigma}_i^{-1} \boldsymbol{\Omega}_i \boldsymbol{\Sigma}_i^{-1} \boldsymbol{D}_i)$ *is the outer product gradient (OPG).*

5. *Strongly consistent estimators of* $\boldsymbol{A}(\boldsymbol{\beta}_0, \boldsymbol{\alpha}_0)$ *and* $\boldsymbol{B}(\boldsymbol{\beta}_0, \boldsymbol{\alpha}_0)$ *are given by*

$$\hat{\boldsymbol{A}} = \frac{1}{n} \sum_{i=1}^{n} \Big(\hat{\boldsymbol{D}}_i' \hat{\boldsymbol{\Sigma}}_i^{-1} \hat{\boldsymbol{D}}_i\Big) \quad \text{and} \quad \hat{\boldsymbol{B}} = \frac{1}{n} \sum_{i=1}^{n} \Big(\hat{\boldsymbol{D}}_i' \hat{\boldsymbol{\Sigma}}_i^{-1} \hat{\boldsymbol{\Omega}}_i \hat{\boldsymbol{\Sigma}}_i^{-1} \hat{\boldsymbol{D}}_i\Big) \,,$$

where $\hat{\boldsymbol{D}}_i = \partial \hat{\boldsymbol{\mu}}_i / \partial \boldsymbol{\beta}'$ *is the estimated matrix of first derivatives of the mean with respect to the parameter vector,* $\hat{\boldsymbol{\Sigma}}_i = \boldsymbol{\Sigma}(\boldsymbol{X}_i, \hat{\boldsymbol{\beta}}, \tilde{\boldsymbol{\Psi}}_i)$ *is the estimator of the covariance matrix of the assumed distribution, and* $\boldsymbol{\Omega}_i$ *is replaced by* $\hat{\boldsymbol{\Omega}}_i = \big(\boldsymbol{y}_i - \hat{\boldsymbol{\mu}}_i\big)\big(\boldsymbol{y}_i - \hat{\boldsymbol{\mu}}_i\big)'$.

6. *The set of asymptotic covariance matrices of the QGPML estimator* $\hat{\boldsymbol{\beta}}$ *of* $\boldsymbol{\beta}$ *based on a linear exponential family has lower bound* $\boldsymbol{\Upsilon}^{-1}(\boldsymbol{\beta}) = \big(\mathbb{E}^{\boldsymbol{X}}(\boldsymbol{D}_i' \boldsymbol{\Omega}_i^{-1} \boldsymbol{D}_i)\big)^{-1}$.

Proof.
1.: Existence: See White (1981, Theorem 2.1).
2.: Consistency: See Gourieroux et al. (1984b, Theorem 4).
3.: Score equations: Analogously to the proof of Theorem 5.2 with $\tilde{\boldsymbol{\Sigma}}_i$ being the replacement of $\boldsymbol{\Sigma}_i$.
5.: Estimation: See White (1982, Theorem 3.2).
6.: Analogously to the proof of Theorem 5.2, 7.
4. Here, we prove the asymptotic normality along the lines of Gourieroux and Monfort (1995a, pp. 215-216, 250). The proof proceeds in three steps. In the first step, we consider a Taylor expansion of the score equations in a neighborhood of $(\boldsymbol{\beta}_0', \boldsymbol{\alpha}_0')'$. In the second step, we formulate a condition for which the extra variability introduced by the estimator $\tilde{\boldsymbol{\alpha}}$ can be neglected. In the final step, we show that the QGPML estimator fulfills this condition.

Step 1: Given the score equations

$$\boldsymbol{u}(\hat{\boldsymbol{\beta}}, \tilde{\boldsymbol{\alpha}}) = \frac{1}{n}\sum_{i=1}^{n} \frac{\partial l(\hat{\boldsymbol{\beta}}, \tilde{\boldsymbol{\alpha}})}{\partial \boldsymbol{\beta}} = \mathbf{0},$$

we obtain

$$\begin{aligned}\mathbf{0} \overset{a.s.}{=} & \frac{1}{\sqrt{n}}\sum_{i=1}^{n} \frac{\partial l_i(\boldsymbol{\beta}_0, \boldsymbol{\alpha}_0)}{\partial \boldsymbol{\beta}} \\ & + \frac{1}{\sqrt{n}}\sum_{i=1}^{n} \frac{\partial^2 l_i(\boldsymbol{\beta}^*, \boldsymbol{\alpha}_0)}{\partial \boldsymbol{\beta}\partial \boldsymbol{\beta}'}(\hat{\boldsymbol{\beta}} - \boldsymbol{\beta}_0) + \frac{1}{\sqrt{n}}\sum_{i=1}^{n} \frac{\partial^2 l_i(\boldsymbol{\beta}_0, \boldsymbol{\alpha}^*)}{\partial \boldsymbol{\beta}\partial \boldsymbol{\alpha}'}(\tilde{\boldsymbol{\alpha}} - \boldsymbol{\alpha}_0)\end{aligned}$$

with a.s. denoting almost surely, and $\boldsymbol{\beta}^*$ and $\boldsymbol{\alpha}^*$ lying on the line segment between $\hat{\boldsymbol{\beta}}$ and $\boldsymbol{\beta}_0$ and $\hat{\boldsymbol{\alpha}}$ and $\boldsymbol{\alpha}$, respectively (see proof of Theorem 4.2). By a strong law of large numbers (White, 1981, Lemma 3.1), we get

$$\begin{aligned}\mathbf{0} \overset{a.s.}{=} & \frac{1}{\sqrt{n}}\sum_{i=1}^{n} \frac{\partial l(\boldsymbol{\beta}_0, \boldsymbol{\alpha}_0)}{\partial \boldsymbol{\beta}} \\ & + \mathbb{E}^{\boldsymbol{X}}\mathbb{E}^{\boldsymbol{y}}\Big(\frac{\partial^2 l_i(\boldsymbol{\beta}_0, \boldsymbol{\alpha}_0)}{\partial \boldsymbol{\beta}\partial \boldsymbol{\beta}'}\Big)(\hat{\boldsymbol{\beta}} - \boldsymbol{\beta}_0) + \mathbb{E}^{\boldsymbol{X}}\mathbb{E}^{\boldsymbol{y}}\Big(\frac{\partial^2 l_i(\boldsymbol{\beta}_0, \boldsymbol{\alpha}_0)}{\partial \boldsymbol{\beta}\partial \boldsymbol{\alpha}'}\Big)(\tilde{\boldsymbol{\alpha}} - \boldsymbol{\alpha}_0).\end{aligned}$$

This results in

$$\sqrt{n}(\hat{\boldsymbol{\beta}} - \boldsymbol{\beta}_0) \overset{a.s.}{=} \boldsymbol{A}^{-1}\Big(\frac{1}{\sqrt{n}}\sum_{i=1}^{n} \frac{\partial l_i(\boldsymbol{\beta}_0, \boldsymbol{\alpha}_0)}{\partial \boldsymbol{\beta}} + \boldsymbol{J}\sqrt{n}(\tilde{\boldsymbol{\alpha}} - \boldsymbol{\alpha}_0)\Big)$$

where

$$\boldsymbol{A}^{-1} = \boldsymbol{A}^{-1}(\boldsymbol{\beta}_0, \boldsymbol{\alpha}_0) = \mathbb{E}^{\boldsymbol{X}}\mathbb{E}^{\boldsymbol{y}}\Big(-\frac{\partial^2 l_i(\boldsymbol{\beta}_0, \boldsymbol{\alpha}_0)}{\partial \boldsymbol{\beta}\partial \boldsymbol{\beta}'}\Big)$$

and

$$\boldsymbol{J} = \boldsymbol{J}(\boldsymbol{\beta}_0, \boldsymbol{\alpha}_0) = \mathbb{E}^{\boldsymbol{X}}\mathbb{E}^{\boldsymbol{y}}\Big(\frac{\partial^2 l_i(\boldsymbol{\beta}_0, \boldsymbol{\alpha}_0)}{\partial \boldsymbol{\beta}\partial \boldsymbol{\alpha}'}\Big).$$

Application of the multivariate central limit theorem (Lehmann and Casella, 1998, p. 61, Theorem 8.21) gives that the vector

$$\begin{pmatrix} \frac{1}{\sqrt{n}}\sum_{i=1}^{n} \frac{\partial l_i(\boldsymbol{\beta}_0, \boldsymbol{\alpha}_0)}{\partial \boldsymbol{\beta}} \\ \sqrt{n}(\tilde{\boldsymbol{\alpha}} - \boldsymbol{\alpha}_0) \end{pmatrix}$$

is asymptotically normally distributed with mean vector $\mathbf{0}$ and covariance matrix

$$\begin{pmatrix} \boldsymbol{B} & \boldsymbol{B}_{0\alpha} \\ \boldsymbol{B}_{\alpha 0} & \boldsymbol{B}_{\alpha\alpha} \end{pmatrix}.$$

Here, the index α indicates the nuisance parameter $\boldsymbol{\alpha}$.

Subsequently, the QGPML estimator $\hat{\boldsymbol{\beta}}$ is asymptotically normally distributed as

$$\sqrt{n}(\hat{\boldsymbol{\beta}} - \boldsymbol{\beta}_0) \overset{a}{\sim} N(\mathbf{0}, \boldsymbol{V}) ,$$

where

$$\boldsymbol{V} = \boldsymbol{A}^{-1}\left((\boldsymbol{I}_{p\times p}, \boldsymbol{J}) \begin{pmatrix} \boldsymbol{B} & \boldsymbol{B}_{0\alpha} \\ \boldsymbol{B}_{\alpha 0} & \boldsymbol{B}_{\alpha\alpha} \end{pmatrix} \begin{pmatrix} \boldsymbol{I}_{p\times p} \\ \boldsymbol{J}' \end{pmatrix} \right) \boldsymbol{A}^{-1} . \tag{6.2}$$

Equation 6.2 shows that the covariance matrix of $\hat{\boldsymbol{\beta}}$ generally differs from the covariance matrix of the PML1 estimator.

In the second step of the proof, we therefore formulate a condition that reduces the covariance matrix to the common robust covariance matrix. Specifically, if

$$\boldsymbol{J} = \boldsymbol{J}(\boldsymbol{\beta}_0, \boldsymbol{\alpha}_0) = \mathbb{E}^{\boldsymbol{X}} \mathbb{E}^{\boldsymbol{y}} \Big(\frac{\partial^2 l_i(\boldsymbol{\beta}_0, \boldsymbol{\alpha}_0)}{\partial \boldsymbol{\beta} \partial \boldsymbol{\alpha}'} \Big) = \mathbf{0} , \tag{6.3}$$

then the QGPML estimator $\hat{\boldsymbol{\beta}}$ is asymptotically normally distributed, more precisely:

$$\sqrt{n}(\hat{\boldsymbol{\beta}} - \boldsymbol{\beta}_0) \overset{a}{\sim} N(\mathbf{0}, \boldsymbol{C}) ,$$

with

$$\begin{aligned}
\boldsymbol{C} &= \boldsymbol{C}(\boldsymbol{\beta}_0, \boldsymbol{\alpha}_0) = \boldsymbol{A}^{-1}(\boldsymbol{\beta}_0, \boldsymbol{\alpha}_0) \boldsymbol{B}(\boldsymbol{\beta}_0, \boldsymbol{\alpha}_0) \boldsymbol{A}^{-1}(\boldsymbol{\beta}_0, \boldsymbol{\alpha}_0) , \\
\boldsymbol{A} &= \boldsymbol{A}(\boldsymbol{\beta}_0, \boldsymbol{\alpha}_0) = \mathbb{E}^{\boldsymbol{X}} \mathbb{E}^{\boldsymbol{y}} \Big(- \frac{\partial^2 l_i(\boldsymbol{\beta}_0, \boldsymbol{\alpha}_0)}{\partial \boldsymbol{\beta} \partial \boldsymbol{\beta}'} \Big) , \\
\boldsymbol{B} &= \boldsymbol{B}(\boldsymbol{\beta}_0, \boldsymbol{\alpha}_0) = \mathbb{E}^{\boldsymbol{X}} \mathbb{E}^{\boldsymbol{y}} \Big(\frac{\partial l_i(\boldsymbol{\beta}_0, \boldsymbol{\alpha}_0)}{\partial \boldsymbol{\beta}} \frac{\partial l_i(\boldsymbol{\beta}_0, \boldsymbol{\alpha}_0)}{\partial \boldsymbol{\beta}'} \Big) .
\end{aligned}$$

In the final step, we need to show that Eq. 6.3 holds for QGPML estimators. This can be done using the formulation of the linear exponential family of Eq. 1.7, the vectorization of the nuisance parameter $\text{vec}(\boldsymbol{\Psi})$, and Property 1.4:

$$\begin{aligned}
\mathbb{E}^{\boldsymbol{y}} \Big(\frac{\partial^2 f(\boldsymbol{y}|\boldsymbol{X}||\boldsymbol{\mu}, \boldsymbol{\Psi})}{\partial \boldsymbol{\mu}\, \partial \text{vec}(\boldsymbol{\Psi})'} \Big) &= \mathbb{E}^{\boldsymbol{y}} \Big(\frac{\partial^2 \mathbf{c}(\boldsymbol{\mu}, \boldsymbol{\Psi})' \boldsymbol{y} + a(\boldsymbol{\mu}, \boldsymbol{\Psi}) + b(\boldsymbol{y}, \boldsymbol{\Psi})}{\partial \boldsymbol{\mu}\, \partial \text{vec}(\boldsymbol{\Psi})'} \Big) \\
&= \mathbb{E}^{\boldsymbol{y}} \Big(\frac{\partial}{\partial \text{vec}(\boldsymbol{\Psi})'} \Big(\frac{\mathbf{c}(\boldsymbol{\mu}, \boldsymbol{\Psi})' \boldsymbol{y}}{\partial \boldsymbol{\mu}} + \frac{\partial a(\boldsymbol{\mu}, \boldsymbol{\Psi})}{\partial \boldsymbol{\mu}} + \frac{\partial b(\boldsymbol{y}, \boldsymbol{\Psi})}{\partial \boldsymbol{\mu}} \Big) \Big) \\
&\overset{(\star)}{=} \frac{\partial}{\partial \text{vec}(\boldsymbol{\Psi})'} \mathbb{E}^{\boldsymbol{y}} \Big(\frac{\mathbf{c}(\boldsymbol{\mu}, \boldsymbol{\Psi})' \boldsymbol{y}}{\partial \boldsymbol{\mu}} + \frac{\partial a(\boldsymbol{\mu}, \boldsymbol{\Psi})}{\partial \boldsymbol{\mu}} \Big) \\
&= \frac{\partial}{\partial \text{vec}(\boldsymbol{\Psi})'} \Big(\frac{\mathbf{c}(\boldsymbol{\mu}, \boldsymbol{\Psi})' \boldsymbol{\mu}}{\partial \boldsymbol{\mu}} + \frac{\partial a(\boldsymbol{\mu}, \boldsymbol{\Psi})}{\partial \boldsymbol{\mu}} \Big) \\
&\overset{(\star\star)}{=} \frac{\partial}{\partial \text{vec}(\boldsymbol{\Psi})'} = \mathbf{0}
\end{aligned}$$

where the interchange of differentiation and integration is used at ($\star$), and Property 1.4 at ($\star\star$). As a consequence, we have shown that Eq. 6.3 is satisfied, and this completes the proof. □

Remark 6.3.

1. Equation 6.3 implies that the extra variability of the nuisance parameter $\boldsymbol{\alpha}$, i.e., the asymptotic distribution of the nuisance parameter estimator $\tilde{\boldsymbol{\alpha}}$, need not be taken into account.
2. So far, we have considered the two-step procedure in the following way. In the first step, the nuisance parameter vector $\boldsymbol{\alpha}_0$ is estimated by $\tilde{\boldsymbol{\alpha}}$ given an initial guess of $\tilde{\boldsymbol{\beta}}$. In the second step, the parameter vector $\boldsymbol{\beta}_0$ is estimated given $\tilde{\boldsymbol{\alpha}}$ and $\tilde{\boldsymbol{\beta}}$. In applications, one should return to step 1 after step 2 and repeat the estimation of $\tilde{\boldsymbol{\alpha}}$ given $\hat{\boldsymbol{\beta}}$ from step 2. This alternating two-step procedure is preferable over the simple two-step approach as shown, e.g., by Carroll and Ruppert (1988). They specifically considered a heteroscedastic linear regression model and demonstrated that several alternating estimation steps were required before the asymptotic covariance matrix of $\hat{\boldsymbol{\beta}}$ stabilized. It can be assumed that similar results hold for general mean structures, and therefore, the two-step procedure sketched above should be iterated until convergence of both $\tilde{\boldsymbol{\alpha}}$ and $\hat{\boldsymbol{\beta}}$.

In Theorem 6.2, we assumed that the true covariance matrix $\boldsymbol{\Omega}_i$ may be misspecified. In the following, we formulate properties of the QGPML estimator for the case that the true covariance matrix $\boldsymbol{\Omega}_i$ and the assumed covariance matrix $\boldsymbol{\Sigma}_i$ are identical. We emphasize that we still do not assume the correct specification of the complete multivariate distribution. Only the correct specification of the first two moments is assumed.

Theorem 6.4 (Asymptotic equivalence of QGPML and ML estimation). *We assume that*

$$\mathbb{V}ar_{f^\star}(\boldsymbol{y}_i|\boldsymbol{X}_i||\boldsymbol{\beta}_0) = \mathbb{V}ar_f(\boldsymbol{y}_i|\boldsymbol{X}_i||\boldsymbol{\beta}_0,\boldsymbol{\alpha}_0) = \boldsymbol{\Omega}(\boldsymbol{X}_i||\boldsymbol{\beta}_0,\boldsymbol{\alpha}_0)$$

for all $i = 1,\dots,n$. Then, the following properties hold:

1. *The Fisher information matrix $\boldsymbol{A}$ equals the outer product gradient $\boldsymbol{B}$ as in the ML situation. Subsequently, $\hat{\boldsymbol{\beta}}$ is distributed as*

$$\sqrt{n}(\hat{\boldsymbol{\beta}} - \boldsymbol{\beta}_0) \overset{a}{\sim} N\big(\mathbf{0}, \big(\boldsymbol{A}(\boldsymbol{\beta}_0,\boldsymbol{\alpha}_0)\big)^{-1}\big) = N\big(\mathbf{0}, \boldsymbol{B}(\boldsymbol{\beta}_0,\boldsymbol{\alpha}_0)^{-1}\big).$$

2. *If both the true and the assumed distribution belong to the linear exponential family, then the QGPML estimator of $\boldsymbol{\beta}_0$ is asymptotically equivalent to the ML estimator of $\boldsymbol{\beta}_0$ obtained by maximizing the true log likelihood with respect to $\boldsymbol{\beta}$ and $\boldsymbol{\alpha}$.*

Proof.
1.: This is a direct consequence of the proof to Theorem 6.2.
2.: See Gourieroux et al. (1984b, Theorem 5). □

Remark 6.5. Theorem 6.4 states that the QGPML estimator asymptotically reaches the lower (Rao–Cramér) bound if both the mean structure and the covariance matrix are correctly specified. The QGPML estimator thus is asymptotically efficient in this case.

6.3 Examples

In this section, a series of different examples for QGPML estimation is given. First, we consider an example for univariate QGPML estimation, i.e., $T = 1$. It has been used in an applied problem on estimating the tree volume of red oak trees (Ziegler et al., 1997), and it is similar to the theoretical example given by Gourieroux et al. (1984b).

Example 6.6 (Estimation of tree volume from breast height diameter). A standard problem of scientific forestry is to provide a prognostic model for tree volumes for wood based on simple measurable quantities. An often used quantity for predicting the tree volume (Vol) is the breast height diameter (BHD), which is the diameter of a tree at a height of 130 cm. In the models, the trees are usually assumed to be independent.

The volume is positive, and therefore, the square root link and the log link are reasonable choices. As an alternative, a linear model for the log tree volume (lnVol) is sometimes considered. A standard feature of tree volume data is that the dispersion increases with BHD. Hence, the choice of a Poisson model for Vol seems to be adequate. However, standard tests for normality typically reject the assumption of normality for lnVol. Summing up, the choice of the loglink function for Vol seems to be justified and preferable over the identity link function for lnVol so that we assume $\mu_i = \mathbb{E}_{f^\star}(y_i|\boldsymbol{x}_i) = \mathbb{E}_f(y_i|\boldsymbol{x}_i) = \exp(\boldsymbol{x}_i'\boldsymbol{\beta})$.

In many applications, a quadratic variance function plus overdispersion is used for the loglink, i.e., $v_i = \mathbb{V}\mathrm{ar}(y_i|\boldsymbol{x}_i) = \Phi \exp^2(\boldsymbol{x}_i'\boldsymbol{\beta})$, where Φ is the dispersion parameter, and $h(\mu_i) = \mu_i^2 = \exp^2(\boldsymbol{x}_i'\boldsymbol{\beta})$ is the variance function. This choice can be justified by the observation that the variance of the BHD approximately follows a quadratic function of the mean. However, the true conditional variance $\Omega_i = \Omega(\boldsymbol{x}_i)$ need not be correctly specified.

In the first step, an initial guess of the parameter $\boldsymbol{\beta}$ is obtained by minimizing the normed sum $\sum_{i=1}^n \big(y_i - \mu(\boldsymbol{x}_i'\boldsymbol{\beta})\big)^2$ with respect to $\boldsymbol{\beta}$. The corresponding estimating equations are

$$\sum_{i=1}^n \big(y_i - \exp(\boldsymbol{x}_i'\tilde{\boldsymbol{\beta}})\big) \frac{\partial \exp(\boldsymbol{x}_i'\tilde{\boldsymbol{\beta}})}{\partial \tilde{\boldsymbol{\beta}}} = \sum_{i=1}^n \big(y_i - \exp(\boldsymbol{x}_i'\tilde{\boldsymbol{\beta}})\big) \exp(\boldsymbol{x}_i'\tilde{\boldsymbol{\beta}})\boldsymbol{x}_i = \boldsymbol{0}\,.$$

This estimating equation is a standard problem of nonlinear optimization (Antoniou and Lu, 2007). With the initial guess $\tilde{\boldsymbol{\beta}}$, a strongly consistent estimator $\tilde{\Phi}$ for Φ can be obtained:

$$\tilde{\Phi} = \frac{1}{n}\sum_{i=1}^{n} \frac{\left(y_i - \exp(\boldsymbol{x}_i'\tilde{\boldsymbol{\beta}})\right)^2}{\exp^2(\boldsymbol{x}_i'\tilde{\boldsymbol{\beta}})} .$$

In the second step, a pseudo likelihood function needs to be specified. For illustration, we first choose the normal distribution. Below, we also illustrate the use of the gamma distribution. To repeat, both the assumed normal distribution and the assumed gamma distribution yield consistent estimates of the parameter vector $\boldsymbol{\beta}$ if the mean structure is correctly specified and if the domain of the mean structure parameter of the true distribution is a subset of the domain of the mean structure parameter of the assumed distribution.

In the first part of this example, the conditional assumed distribution of y_i given $\boldsymbol{x}_i$ is the normal distribution with conditional mean $\mu(\boldsymbol{x}_i'\boldsymbol{\beta})$ and fixed conditional variance $\tilde{v}_i$. The kernel of the normed pseudo loglikelihood is given by

$$l(\boldsymbol{\beta}) = \frac{1}{n}\sum_{i=1}^{n} -\frac{1}{2}\frac{\left(y_i - \exp(\boldsymbol{x}_i'\boldsymbol{\beta})\right)^2}{\tilde{\Phi}\exp^2(\boldsymbol{x}_i'\tilde{\boldsymbol{\beta}})} = \frac{1}{n}\sum_{i=1}^{n} -\frac{1}{2}\frac{\left(y_i - \exp(\boldsymbol{x}_i'\boldsymbol{\beta})\right)^2}{\tilde{v}_i} ,$$

which has to be maximized with respect to $\boldsymbol{\beta}$. Alternatively, the negative of this expression is minimized with respect to $\boldsymbol{\beta}$. The first and second derivatives of the kernel of the negative individual pseudo loglikelihood are given by

$$\boldsymbol{u}_i(\boldsymbol{\beta}) = \frac{\partial}{\partial\boldsymbol{\beta}}\frac{\left((y_i - \exp(\boldsymbol{x}_i'\boldsymbol{\beta})\right)^2}{2\,\tilde{v}_i} = -\frac{1}{\tilde{v}_i}\mu_i(y_i - \mu_i)\boldsymbol{x}_i = -\frac{\mu_i}{\tilde{v}_i}(y_i - \mu_i)\boldsymbol{x}_i$$

and

$$\boldsymbol{W}_i(\boldsymbol{\beta}) = \frac{\partial \boldsymbol{u}_i(\boldsymbol{\beta})}{\partial\boldsymbol{\beta}'} = -\frac{1}{\tilde{v}_i}\mu_i(y_i - 2\mu_i)\boldsymbol{x}_i\boldsymbol{x}_i' .$$

For non-stochastic explanatory variables, the expected values of the individual OPG and the individual Fisher information matrix are

$$\mathbb{E}(\boldsymbol{u}_i\boldsymbol{u}_i') = \frac{\mu_i^2}{v_i^2}\Omega_i\boldsymbol{x}_i\boldsymbol{x}_i' \quad \text{and} \quad \mathbb{E}(\boldsymbol{W}_i) = \frac{\mu_i^2}{v_i}\boldsymbol{x}_i\boldsymbol{x}_i' .$$

The individual OPG equals the individual Fisher information matrix if $\Omega_i = v_i$, i.e., if the assumed variance v_i equals the true variance Ω_i. With the quantities derived, $\hat{\boldsymbol{\beta}}$ can be estimated by a modified Fisher scoring algorithm with fixed $\tilde{v}_i$. For stabilization of parameter estimates, both steps are iterated until convergence.

After convergence, the variance of $\hat{\boldsymbol{\beta}}$ is estimated by

$$\widehat{\mathbb{V}\text{ar}}(\hat{\boldsymbol{\beta}}) = \left(\boldsymbol{X}'\text{diag}\Big(\frac{\hat{\mu}_i^2}{\tilde{v}_i}\Big)\boldsymbol{X}\right)^{-1}\left(\boldsymbol{X}'\text{diag}\Big(\frac{\hat{\mu}_i^2(y_i - \hat{\mu}_i)^2}{\tilde{v}_i^2}\Big)\boldsymbol{X}\right)\left(\boldsymbol{X}'\text{diag}\Big(\frac{\hat{\mu}_i^2}{\tilde{v}_i}\Big)\boldsymbol{X}\right)^{-1} .$$

In the second part of this example, the conditional assumed distribution of y_i given $\boldsymbol{x}_i$ is the gamma distribution. The kernel of the normed loglikelihood function of the gamma distribution is given by (see Example 1.11)

$$l(\boldsymbol{\beta}) = \frac{1}{n}\sum_{i=1}^{n}\left(\Psi_i \ln\frac{\Psi_i}{\mu_i} - \frac{\Psi_i}{\mu_i}y_i\right).$$

The term $\Psi_i \ln \Psi_i$ is irrelevant for maximization, and given an estimate $\tilde{\Psi}_i$, the kernel can be written either as

$$\frac{1}{n}\sum_{i=1}^{n} -\tilde{\Psi}_i\Big(\ln\exp(\boldsymbol{x}_i'\boldsymbol{\beta}) + \frac{y_i}{\exp(\boldsymbol{x}_i'\boldsymbol{\beta})}\Big)$$

or equivalently as

$$\frac{1}{n}\sum_{i=1}^{n} -\frac{\exp^2(\boldsymbol{x}_i'\tilde{\boldsymbol{\beta}})}{\tilde{\Phi}\exp^2(\boldsymbol{x}_i\tilde{\boldsymbol{\beta}})}\Big(\ln\exp(\boldsymbol{x}_i'\boldsymbol{\beta}) + \frac{y_i}{\exp(\boldsymbol{x}_i'\boldsymbol{\beta})}\Big) =$$
$$\frac{1}{n}\sum_{i=1}^{n} -\frac{1}{\tilde{\Phi}}\Big(\ln\exp(\boldsymbol{x}_i'\boldsymbol{\beta}) + \frac{y_i}{\exp(\boldsymbol{x}_i'\boldsymbol{\beta})}\Big).$$

The first and second derivatives of the kernel of the negative individual pseudo loglikelihood are given by

$$\boldsymbol{u}_i(\boldsymbol{\beta}) = \frac{1}{\tilde{\Phi}}\Big(\frac{1}{\exp(\boldsymbol{x}_i'\boldsymbol{\beta})} - \frac{y_i}{\exp^2(\boldsymbol{x}_i'\boldsymbol{\beta})}\Big)\boldsymbol{x}_i = -\frac{1}{\tilde{\Phi}}\frac{y_i-\mu_i}{\mu_i^2}\boldsymbol{x}_i$$

and

$$\boldsymbol{W}_i(\boldsymbol{\beta}) = -\frac{1}{\tilde{\Phi}}\Big(\frac{1}{\mu_i^2} - \frac{2y_i}{\mu_i^3}\Big)\boldsymbol{x}_i\boldsymbol{x}_i' = \frac{1}{\tilde{\Phi}}\frac{2y_i-\mu_i}{\mu_i^3}\boldsymbol{x}_i\boldsymbol{x}_i'.$$

For non-stochastic explanatory variables, the expected values of the individual OPG and the individual Fisher information matrix are

$$\mathbb{E}(\boldsymbol{u}_i\boldsymbol{u}_i') = \frac{1}{\Phi^2\mu_i^4}\Omega_i\boldsymbol{x}_i\boldsymbol{x}_i' \quad\text{and}\quad \mathbb{E}(\boldsymbol{W}_i) = \frac{1}{\Phi\mu_i^2}\boldsymbol{x}_i\boldsymbol{x}_i'.$$

The individual OPG equals the individual Fisher information matrix if $\Omega_i = \Phi\mu_i^2 = v_i$, i.e., if the assumed variance v_i equals the true variance Ω_i. The maximization problems can be solved by a modified Fisher scoring algorithm with fixed $\tilde{\Phi}$. For stabilization of parameter estimates, both steps are iterated until convergence.

After convergence, the variance of $\hat{\boldsymbol{\beta}}$ is estimated by

$$\widehat{\mathbb{V}\mathrm{ar}}(\hat{\boldsymbol{\beta}}) = \Big(\boldsymbol{X}'\mathrm{diag}\Big(\frac{1}{\tilde{\Phi}\hat{\mu}_i^2}\Big)\boldsymbol{X}\Big)^{-1}\Big(\boldsymbol{X}'\mathrm{diag}\Big(\frac{(y_i-\hat{\mu}_i)^2}{\tilde{\Phi}^2\hat{\mu}_i^4}\Big)\boldsymbol{X}\Big)\Big(\boldsymbol{X}'\mathrm{diag}\Big(\frac{1}{\tilde{\Phi}\hat{\mu}_i^2}\Big)\boldsymbol{X}\Big)^{-1}.$$

To repeat, the QGPML estimators based on the normal distribution and the gamma distribution are asymptotically equivalent.

In all following examples, we consider clustered data, more specifically, we consider n independently distributed T-dimensional random vectors $\boldsymbol{y}_i$, $i = 1, \ldots, n$, and $\boldsymbol{X}_i$ is the $T \times p$ matrix of fixed explanatory variables of subject i.

6.3.1 Generalized estimating equations 1 with estimated working covariance matrix

In Sect. 5.3.5, we considered the GEE1 with a fixed covariance matrix using PML1 estimation. In this section, we generalize the results from Sect. 5.3.5 and allow for an estimated working covariance matrix. This generalization has one obvious advantage. In most applications, no a priori information is available, how specific values of the covariance matrix should be chosen, although information on a reasonable covariance structure might be available. With the QGPML approach, efficiency of parameter estimates might be improved by first assuming a reasonable working covariance structure, second, estimating this working covariance structure, and finally, estimating the parameters of interest of the mean structure.

For illustration, we consider the exchangeable covariance structure $\boldsymbol{\Sigma}_i = \boldsymbol{\Sigma}$ with entries

$$\mathbb{V}\text{ar}(y_{it}) = \sigma^{(1)} \quad \text{and} \quad \mathbb{C}\text{ov}(y_{it}, y_{it'}) = \sigma^{(12)} ,$$

for $t, t' = 1, \ldots, T$ and $i = 1, \ldots, n$. Application of QGPML estimation proceeds as follows:

1. An estimate $\tilde{\boldsymbol{\beta}}$ of $\boldsymbol{\beta}$ is obtained, e.g., using the assumption of independence, i.e., by minimizing $\frac{1}{n}\sum_{i=1}^{n}(\boldsymbol{y}_i - \boldsymbol{\mu}_i)'(\boldsymbol{y}_i - \boldsymbol{\mu}_i)$. As a result, the time-point-specific variances σ_t^2 can be estimated by

$$\tilde{\sigma}_t^2 = \frac{1}{n}\sum_{i=1}^{n}(y_{it} - \tilde{\mu}_{it})^2 ,$$

and the time-point-specific covariances $\sigma_{tt'}$ are estimated by

$$\tilde{\sigma}_{tt'} = \frac{1}{n}\sum_{i=1}^{n}(y_{it} - \tilde{\mu}_{it})(y_{it'} - \tilde{\mu}_{it'}) ,$$

where $\tilde{\mu}_{it} = g(\boldsymbol{x}_{it}'\tilde{\boldsymbol{\beta}})$ as in GLM.
Using the structure of the covariance matrix, estimates of $\sigma^{(1)}$ and $\sigma^{(12)}$ can be obtained by

$$\tilde{\sigma}^{(1)} = \frac{1}{T}\sum_{t=1}^{T}\tilde{\sigma}^2 \quad \text{and} \quad \tilde{\sigma}^{(12)} = \frac{2}{T(T-1)}\sum_{t>t'}\tilde{\sigma}_{tt'} .$$

2. $\tilde{\boldsymbol{\Sigma}}$ is considered fixed and used as the conditional variance matrix of the assumed distribution. The distributional assumption for PML1 estimation with fixed $\tilde{\boldsymbol{\Sigma}}$ is $\boldsymbol{y}_i|\boldsymbol{X}_i \sim N(\boldsymbol{\mu}_i, \tilde{\boldsymbol{\Sigma}})$.
   The kernel of the individual pseudo loglikelihood function is given by

$$l_i(\boldsymbol{\beta}|\tilde{\boldsymbol{\Sigma}}) = -\frac{1}{2}(\boldsymbol{y}_i - \boldsymbol{\mu}_i)'\tilde{\boldsymbol{\Sigma}}^{-1}(\boldsymbol{y}_i - \boldsymbol{\mu}_i)$$

   after the addition of $-\frac{1}{2}\boldsymbol{y}_i'\tilde{\boldsymbol{\Sigma}}^{-1}\boldsymbol{y}_i$. The resulting estimating equations are given by

$$\boldsymbol{u}(\hat{\boldsymbol{\beta}}) = \frac{1}{n}\sum_{i=1}^{n}\hat{\boldsymbol{D}}_i'\tilde{\boldsymbol{\Sigma}}^{-1}\hat{\varepsilon}_i = \boldsymbol{0} .$$

   These estimating equations are similar to those of Section 5.3.5. However, $\tilde{\boldsymbol{\Sigma}}$ is used instead of $\boldsymbol{\Sigma}$.

6.3.2 Independence estimating equations

A disadvantage of the estimating equations considered in the previous section is that information of the variance function $h(\mu_{it})$ from GLM is ignored. In this section, we consider estimating equations, where this specific functional relationship between the mean and the variance is taken into account. The starting point for these estimating equations, which are termed independence estimating equations (IEE), is the mean structure model from a GLM assuming independence:

$$\mathbb{E}(y_{it}|\boldsymbol{x}_{it}) = \mathbb{E}(y_{it}|\boldsymbol{X}_i) = g(\boldsymbol{x}_{it}'\boldsymbol{\beta}) .$$

Similarly, we use the variance from a GLM:

$$\mathbb{V}\text{ar}(y_{it}|\boldsymbol{x}_{it}) = v_{it} = \Psi\, h(\mu_{it}) .$$

For simplicity, we assume that y_{it} and $y_{it'}$ are pairwise uncorrelated so that $\mathbb{C}\text{ov}(y_{it}, y_{it'}) = 0$ if $t \neq t'$. As before, the true covariance matrix is $\boldsymbol{\Omega}_i$. For estimation, we use the normal distribution as assumed distribution, i.e., $\boldsymbol{y}_i \sim N(\boldsymbol{\mu}_i, \boldsymbol{\Sigma}_i)$, where $\boldsymbol{\mu}_i = (\mu_{i1}, \dots, \mu_{iT})'$ and $\boldsymbol{\Sigma}_i = \text{diag}(v_{it})$.

In the first step, we estimate $\tilde{\boldsymbol{\beta}}$ from $\frac{1}{n}\sum_{i=1}^{n}(\boldsymbol{y}_i - \boldsymbol{\mu}_i)'(\boldsymbol{y}_i - \boldsymbol{\mu}_i)$ using nonlinear optimization. Given $\tilde{\boldsymbol{\beta}}$ we fix $h(\tilde{\mu}_{it})$. Next, we estimate the scale parameter Ψ through

$$\tilde{\Psi} = \frac{1}{nT}\sum_{i=1}^{n}\sum_{t=1}^{T}\frac{(y_{it}-\tilde{\mu}_{it})^2}{h(\tilde{\mu}_{it})}.$$

Given $\tilde{v}_{it} = \tilde{\Psi}\, h(\tilde{\mu}_{it})$, we consider $\tilde{\boldsymbol{\Sigma}}_i = \text{diag}(\tilde{v}_{it})$ fixed and use the normal distribution $\boldsymbol{y}_i|\boldsymbol{X}_i \sim N(\boldsymbol{\mu}_i, \tilde{\boldsymbol{\Sigma}}_i)$ as assumed distribution.

The kernel of the individual pseudo loglikelihood function is given by

$$l_i(\boldsymbol{\beta}|\tilde{\boldsymbol{\Sigma}}) = -\frac{1}{2}(\boldsymbol{y}_i - \boldsymbol{\mu}_i)'\tilde{\boldsymbol{\Sigma}}_i^{-1}(\boldsymbol{y}_i - \boldsymbol{\mu}_i) = -\frac{1}{2}(\boldsymbol{y}_i - \boldsymbol{\mu}_i)'\text{diag}(\tilde{v}_{it}^{-1})(\boldsymbol{y}_i - \boldsymbol{\mu}_i)\,,$$

and one can solve the IEE using nonlinear optimization in the second step of QGPML estimation through

$$\boldsymbol{u}(\hat{\boldsymbol{\beta}}) = \frac{1}{n}\sum_{i=1}^{n}\hat{\boldsymbol{D}}_i'\tilde{\boldsymbol{\Sigma}}_i^{-1}\hat{\boldsymbol{\varepsilon}}_i = \frac{1}{n}\sum_{i=1}^{n}\hat{\boldsymbol{D}}_i'\text{diag}(\tilde{v}_{it}^{-1})\hat{\boldsymbol{\varepsilon}}_i = \boldsymbol{0}\,,$$

with $\tilde{\boldsymbol{\Sigma}}_i = \text{diag}(\tilde{v}_{it})$.

6.3.3 Generalized estimating equations 1 with estimated working correlation matrix

In the last section, the IEE were considered using estimated working variances. They were originally proposed by Zeger et al. (1985), and they can be considered the precursor of the GEE1, which were published 1 year later in two different well-recognized articles (Liang and Zeger, 1986; Zeger and Liang, 1986). The fundamental idea of Liang and Zeger was to overcome the possible inefficiency of the IEE. Indeed, for the IEE, $\mathbb{C}\text{ov}(y_{it}, y_{it'}) = \mathbb{C}\text{orr}(y_{it}, y_{it'}) = 0$ for $t \neq t'$ is assumed, which can lead to a substantial loss of efficiency if the true correlation matrix is different from a diagonal matrix.

Liang and Zeger combined the functional structure from GLM with an assumed correlation matrix. In detail, they used the mean structure and variance function from a GLM:

$$\mathbb{E}(y_{it}|\boldsymbol{x}_{it}) = \mathbb{E}(y_{it}|\boldsymbol{X}_i) = g(\boldsymbol{x}_{it}'\boldsymbol{\beta}) \quad \text{and} \quad \mathbb{V}\text{ar}(y_{it}|\boldsymbol{x}_{it}) = v_{it} = \Psi\, h(\mu_{it})\,.$$

With the functional relationship

$$\boldsymbol{\Sigma}_i(\boldsymbol{\beta}, \boldsymbol{\alpha}, \Psi) = \boldsymbol{\Sigma}_i = \boldsymbol{V}_i^{1/2}\boldsymbol{R}_i(\boldsymbol{\alpha})\boldsymbol{V}_i^{1/2}$$

given $\boldsymbol{V}_i = \boldsymbol{V}_i(\boldsymbol{\beta}, \Psi) = \text{diag}(v_{it})$, they introduced a working correlation matrix $\boldsymbol{R}_i(\boldsymbol{\alpha})$, which may depend on a q dimensional nuisance parameter vector $\boldsymbol{\alpha}$ but which is independent of the mean structure parameter $\boldsymbol{\beta}$, i.e., orthogonal to $\boldsymbol{\beta}$ (Cox and Reid, 1987). Specific choices of the working correlation matrix are considered in the next section.

Generally, the index i is omitted, and a single working correlation matrix $\boldsymbol{R}(\boldsymbol{\alpha}) = \boldsymbol{R}_i(\boldsymbol{\alpha})$ is used for all clusters i. In this case, $\boldsymbol{\Sigma}_i(\boldsymbol{\beta},\boldsymbol{\alpha}) = \boldsymbol{V}_i^{1/2}(\boldsymbol{\beta},\Psi)\boldsymbol{R}(\boldsymbol{\alpha})\boldsymbol{V}_i^{1/2}(\boldsymbol{\beta},\Psi)$ is the working variance matrix.

As for all QGPML approaches, estimation proceeds in two steps. First, an initial guess of $\tilde{\boldsymbol{\beta}}$ is obtained. This can be done, e.g., by employing a standard GLM, where the correlation between observations within a cluster is neglected. This approach can also be used for obtaining $\tilde{\Psi}$, an estimate of Ψ. Second, the structural parameter vector $\tilde{\boldsymbol{\alpha}}$ of the working correlation structure is estimated.

The estimates $\tilde{\boldsymbol{\alpha}}$, $\tilde{\boldsymbol{\beta}}$, and $\tilde{\Psi}$ determine the working covariance matrices $\tilde{\boldsymbol{\Sigma}}_i$ for all i, which is considered fixed for the second step of QGPML estimation. A multivariate normal distribution is chosen as assumed distribution for $\boldsymbol{y}_i$ given $\boldsymbol{X}_i$:

$$\boldsymbol{y}_i|\boldsymbol{X}_i \sim N(\boldsymbol{\mu}_i, \tilde{\boldsymbol{\Sigma}}_i)\,.$$

The kernel of an individual (pseudo) loglikelihood function is given by

$$l_i(\boldsymbol{\beta}|\tilde{\boldsymbol{\Sigma}}_i) = -\frac{1}{2}(\boldsymbol{y}_i - \boldsymbol{\mu}_i)\tilde{\boldsymbol{\Sigma}}_i^{-1}(\boldsymbol{y}_i - \boldsymbol{\mu}_i)\,.$$

The resulting estimating equations are obtained by differentiating the normed pseudo loglikelihood function with respect to $\boldsymbol{\beta}$. They are termed generalized estimating equations of order 1, and they are given by

$$\boldsymbol{u}(\hat{\boldsymbol{\beta}}) = \frac{1}{n}\sum_{i=1}^{n} \hat{\boldsymbol{D}}_i'\tilde{\boldsymbol{\Sigma}}_i^{-1}\hat{\boldsymbol{\varepsilon}}_i = \boldsymbol{0}\,. \tag{6.4}$$

These estimating equations slightly differ from the GEE proposed by Liang and Zeger (1986). Specifically, Liang and Zeger (1986) did not fix the complete working covariance matrix in their estimating equations using $\tilde{\boldsymbol{\alpha}}$, $\tilde{\boldsymbol{\beta}}$, and $\tilde{\Psi}$. Instead, they fixed only $\tilde{\boldsymbol{\alpha}}$ and $\tilde{\Psi}$. Subsequently, they allowed $\boldsymbol{\beta}$ to vary in the variance function $h(\mu_{it})$. This difference is negligible for applications because the two steps from QGPML estimation are repeated until convergence.

A remark has to be made about the correct specification of the mean structure at this point. The specification of the mean structure consists of two parts, the choice of the link function g and linear combination of the independent variables $\boldsymbol{x}_i'\boldsymbol{\beta}$. For GEE, the assumption that the mean structure needs to be correctly specified can be weakened as shown by Park and Weisberg (1998). They showed that under common circumstances, consistent estimates of regression coefficients are obtained even if the link function in the generalized linear model is misspecified. The misspecification of the link function can be tested using a goodness-of-link test as described by Molefe and Hosmane (2007).

6.3.4 Examples for working covariance and correlation structures

In this section, we consider common choices for working correlation matrices. Let the working correlation between subjects t and t' of cluster i be $\varrho_{itt'} = \mathbb{C}orr(y_{it}, y_{it'})$. The elements $\varrho_{itt'}$ are summarized to $\boldsymbol{R}_i = \mathbb{C}orr(\boldsymbol{y}_i|\boldsymbol{X}_i)$. Finally, we note that the assumption of independence of clusters implies $\mathbb{C}orr(y_{it}, y_{jt'}) = 0$ for $i \neq j$.

Standard choices for working correlation matrices in standard software packages are (Ziegler, 2012)

- fixed,
- independent,
- exchangeable,
- m-dependent,
- autoregressive, and
- unstructured.

Below, we also consider several non-standard working correlation structures.

Example 6.7 (Fixed working correlation structure). A simple but rarely used working correlation structure is the fixed working correlation structure, which is also termed the user-defined working correlation structure (common abbreviations: FIX, FIXED, USER). Here, the researcher pre-specifies not only the structure of the working correlation matrix but also all values of the working correlation matrix.

If the fixed working correlation structure is used and if no scale parameter Ψ of the variance v_{it} is included in the model, the estimating equations can be solved by PML1 estimation, and they do not require QGPML (Sect. 5.3.5).

Example 6.8 (Independent working correlation structure). The IEE from Sect. 6.3.2 represent a special case of the more general GEE1 (Eq. 6.4) by letting $\boldsymbol{R}_i(\boldsymbol{\alpha})$ equal the identity matrix. The working correlation structure is the independent working correlation structure (common abbreviations: IND, INDE, INDEP). Here,

$$\mathbb{C}orr(y_{it}, y_{it'}) = \begin{cases} 1, & \text{if } t = t'\,, \\ 0, & \text{if } t \neq t'\,. \end{cases}$$

No correlation parameter needs to be estimated in this case.

Example 6.9 (Exchangeable working correlation structure). The exchangeable working correlation structure, also termed the compound symmetry working correlation structure, is a natural choice in family studies and household studies, i.e., in the case of cluster sampling (common abbreviations: EX, EXCH, CS). It is given by

$$\mathbb{C}orr(y_{it}, y_{it'}) = \begin{cases} 1, & \text{if } t = t' , \\ \varrho, & \text{if } t \neq t' . \end{cases}$$

The number of parameters to be estimated is 1.

Although this working correlation structure assumes that the correlation between all observations within a cluster is equal, it generally is a good choice even if the true correlation differs slightly between observations in a cluster. It is often also appropriate if the correlation varies between clusters, e.g., if two different treatments, say treat_1 and treat_2, are applied, leading to correlations ϱ_1 and ϱ_2.

Example 6.10 (Stationary working correlation structure). A working correlation structure that might be of interest for use with longitudinal data is the stationary working correlation structure (common abbreviations: STA, STAT). Here, all measurements with a specific distance in time have equal correlations, and the general definition of the stationary working correlation is

$$\mathbb{C}orr(y_{it}, y_{it'}) = \begin{cases} 1, & \text{if } t = t' , \\ \varrho_{|t-t'|}, & \text{if } t \neq t' . \end{cases}$$

The number of parameters to be estimated is $T - 1$.

Example 6.11 (m-dependent stationary working correlation structure).

The m-dependent stationary working correlation structure is a simplification of the stationary working correlation structure (common abbreviation: MDEP(m), where m is a number for the depth). The assumption is that there is a band of stationary correlations such that all correlations are truncated to zero after the mth band. This does not adequately reflect biological structures over time, and it therefore is not often used in applications. The definition of the m-dependent correlation is

$$\mathbb{C}orr(y_{it}, y_{it'}) = \begin{cases} 1, & \text{if } t = t' , \\ \varrho_{t-t'}, & \text{if } t \neq t' \text{ and } |t - t'| \leq m , \\ 0, & \text{if } |t - t'| > m . \end{cases}$$

The number of parameters to be estimated equals the band width m. If $m = T - 1$, the m-dependent stationary working correlation structure equals the stationary working correlation structure.

Example 6.12 (m-dependent non-stationary working correlation structure). A generalization of the m-dependent working correlation structure is the m-dependent non-stationary working correlation structure, which is given by

$$\mathbb{C}\text{orr}(y_{it}, y_{i,t'}) = \begin{cases} 1 & \text{if } t = t' , \\ \varrho_{t,s} & \text{if } |t - t'| = s \leq m , \\ 0 & \text{if } |t - t'| > m . \end{cases}$$

The number of parameters to be estimated equals $\sum_{l=1}^{m}(T-l)$ and depends on both the band width and the cluster size.

This working correlation structure is rarely used in applications because it does not adequately reflect the biological nature of the data.

Example 6.13 (Autoregressive working correlation structure). A more reasonable working correlation structure for repeated measurements than the m-dependent working correlation structures is the autoregressive working correlation of order 1 (common abbreviations: AR, AR(1)). It is given by

$$\mathbb{C}orr(y_{it}, y_{it'}) = \begin{cases} 1, & \text{if } t = t', \\ \varrho^{|t-t'|}, & \text{if } t \neq t'. \end{cases}$$

The number of parameters to be estimated is 1.

This working correlation structure is often used in applications. It reflects that all observations are correlated but with an exponential decay of the correlation over time.

Example 6.14 (m-dependent autoregressive working correlation structure). In the m-dependent autoregressive working correlation, a band is introduced in analogy to the m-dependent stationary working correlation structure. It is given by

$$\mathbb{C}orr(y_{it}, y_{it'}) = \begin{cases} 1, & \text{if } t = t', \\ \varrho^{|t-t'|}, & \text{if } t \neq t' \text{ and } |t-t'| \leq m, \\ 0, & \text{if } |t-t'| > m. \end{cases}$$

The number of parameters to be estimated is 1. The m-dependent autoregressive working correlation structure equals the AR(1) structure if $m = T - 1$.

It is not often used in applications because it does not reflect the biological nature of the data.

Example 6.15 (Combination of exchangeable and AR(1) working correlation structure). In econometric applications, the working structure sometimes is a combination of the exchangeable and the AR(1) structure, and the variances and covariances of this combination are

$$\sigma_{tt'} = \begin{cases} \sigma_\alpha^2 + \frac{\sigma_\gamma^2}{1-\varrho^2} & \text{if } t = t', \\ \sigma_\alpha^2 + \frac{\sigma_\gamma^2}{1-\varrho^2}\varrho^{|t-t'|} & \text{if } t \neq t'. \end{cases}$$

Here, σ_α^2 is the variance of a random effects model, ϱ is the correlation of y_{it} and $y_{it'}$, and σ_γ^2 reflects the variance of an AR(1) process. As a result, the following working correlation structure is obtained for $t \neq t'$:

$$\varrho_{tt'} = \alpha_1 + \alpha_2 \varrho^{|t-t'|}.$$

Estimation of this working correlation structure requires nonlinear optimization. This working correlation structure is therefore not available as a standard option in common GEE software packages.

Example 6.16 (Unstructured working correlation). The final common working correlation does not make any assumption on a specific structure, and it is therefore called the unstructured working correlation structure (common abbreviations: UN, UNSTR). It is defined as

$$\mathbb{C}orr(y_{it}, y_{it'}) = \begin{cases} 1, & \text{if } t = t' , \\ \varrho_{tt'}, & \text{if } t \neq t' . \end{cases}$$

The number of parameters to be estimated is $T(T-1)/2$, and it may therefore suffer from instability (for a recent example, see, Ziegler and Vens, 2010). It is generally useful only if there is a natural ordering of the observations within a cluster. Furthermore, cluster sizes should be similar because it is not reasonable to estimate a correlation coefficient from one or two pairs of observations.

Example 6.17 (Spatial correlation structure). In most applications, correlations between sampling units are positive (see, e.g., Ziegler, 2012). Examples of negative correlations are uncommon but they deserve special attention. In forest damage surveys, the state of a tree, e.g., the degree of defoliation, is measured. A typical survey uses a grid with rectangular meshes over a map in the survey area. For each grid point, the damage is measured for a fixed number of trees next to the grid point. In this example, a spatial periodic variation can be observed (Baradat et al., 1996), and in some examples, it might be described by simple sine curves (Cochran, 1963, pp. 218-219). As a result, a reasonable working correlation structure in this case could be

$$\mathbb{C}orr(y_{it}, y_{it'}) = \begin{cases} 1, & \text{if } t = t' , \\ (-1)^{t-t'} \varrho^{|t-t'|}, & \text{if } t \neq t' , \end{cases}$$

or

$$\mathbb{C}orr(y_{it}, y_{it'}) = \begin{cases} 1, & \text{if } t = t' , \\ (-1)^{t-t'} \dfrac{\sin\left(2\pi \frac{|t-t'|}{T-1}\right)}{2\pi \frac{|t-t'|}{T-1}}, & \text{if } t \neq t' . \end{cases}$$

The latter example is only reasonable for larger sine waves, say $T \geq 10$, while the first example is also reasonable for shorter waves, i.e., $T \leq 5$.

The choice of the working correlation matrix has been discussed in several articles, and the reader may refer to the literature (Hin and Wang, 2009; Molenberghs, 2010; Sabo and Chaganty, 2010; Shults et al., 2009; Shults, 2011; Ziegler and Vens, 2010).

6.4 Generalizations

In this section, two generalizations of QGPML are considered that are of great importance for GEE1. First, we discuss the analysis of time dependent parameters, and second, we extend the QGPML approach to ordinal dependent variables.

6.4.1 Time dependent parameters

In all previous sections, we implicitly assumed that the parameter vector of interest is constant over the different time points t. The extension to time-varying parameters has been discussed in several articles (see, e.g., Davis, 1991; Schildcrout and Heagerty, 2005; Stram et al., 1988; Wei and Stram, 1988).

We start by considering the mean and variance of y_{it} given $\boldsymbol{x}_{it}$:

$$\mathbb{E}(y_{it}|\boldsymbol{x}_{it}) = \mu_{it} = g(\boldsymbol{x}_{it}'\boldsymbol{\beta}_t) \quad \text{and} \quad \mathbb{V}\text{ar}(y_{it}|\boldsymbol{x}_{it}) = \Psi_t\, h(\mu_{it})\,.$$

This generalization implies that the first column vectors $\boldsymbol{x}_{i1}, \ldots, \boldsymbol{x}_{iT}$ are not stacked to a $T \times p$ matrix $\boldsymbol{X}_i = (\boldsymbol{x}_{i1}', \ldots, \boldsymbol{x}_{iT}')'$ for the analysis of time-varying parameters. One forms the $pT\times$ matrix

$$\boldsymbol{X}_i^* = \begin{pmatrix} \boldsymbol{x}_{i1}' & \boldsymbol{0} & \cdots & \boldsymbol{0} \\ \boldsymbol{0} & \boldsymbol{x}_{i2}' & \ddots & \vdots \\ \vdots & \ddots & \ddots & \boldsymbol{0} \\ \boldsymbol{0} & \cdots & \boldsymbol{0} & \boldsymbol{x}_{iT}' \end{pmatrix} = \boldsymbol{X}_i \otimes \boldsymbol{I}_T\,,$$

and the pT dimensional parameter vector $\boldsymbol{\beta} = (\boldsymbol{\beta}_1', \ldots, \boldsymbol{\beta}_T')'$ instead. Here, $\otimes$ is the Kronecker product and $\boldsymbol{I}_T$ denotes the $T \times T$ identity matrix. As a result,

$$\boldsymbol{X}_i\boldsymbol{\beta} = \begin{pmatrix} \boldsymbol{x}_{i1}'\boldsymbol{\beta}_1 \\ \vdots \\ \boldsymbol{x}_{iT}'\boldsymbol{\beta}_T \end{pmatrix}.$$

Because time-point-specific measurements are independent, time-point-specific estimating equations are given by

$$\boldsymbol{u}(\hat{\boldsymbol{\beta}}_t) = \frac{1}{n}\sum_{i=1}^{n} \frac{\partial \hat{\mu}_{it}}{\partial \boldsymbol{\beta}_t} \tilde{v}_{it}^{-1}(y_{it} - \hat{\mu}_{it}) = \boldsymbol{0}\,.$$

For the sake of simplicity, an independence working correlation matrix is used for different time points. The vector $\hat{\boldsymbol{\beta}} = (\hat{\boldsymbol{\beta}}_1', \ldots, \hat{\boldsymbol{\beta}}_T')'$ is jointly asymptotically normally distributed, and the estimator of the covariance matrix has been given, e.g., by Wei and Stram (1988).

The use of this extension in one of the standard GEE programs is simple. They require only specification of the design matrix $\boldsymbol{X}$ and the vector of dependent variables $\boldsymbol{y}$. The vector of dependent variables is not altered. Finally, $\boldsymbol{X}_i^*$ is obtained by using the Kronecker product as described above.

In many studies with repeated measurements, it is of interest whether parameters are time dependent or constant over time. Such restrictions can be tested using, e.g., the minimum distance estimation (MDE) approach (Sect. 4.3). It is, however, important to note that for dichotomous dependent variables, only proportionality of parameters can be tested, and not the equality of parameters.

This phenomenon is best explained using the univariate threshold model of Sect. 3.1.4. We have already seen that two restrictions had to be introduced for guaranteeing identifiability of parameters. Specifically, the threshold parameter τ and the variance σ^2 were set to 0 and 1, respectively. As a result, the parameter $\boldsymbol{\beta}$ is identifiable only up to scale, and only the latent parameters $\boldsymbol{\beta}_t^* = \boldsymbol{\beta}_t/\sigma_t$ are identified in the longitudinal setting. Therefore, the equality hypothesis

$$H_0 : \boldsymbol{\beta}_1 = \ldots = \boldsymbol{\beta}_T$$

cannot be tested but the hypothesis of proportionality can, i.e.,

$$H_0^* : \boldsymbol{\beta}_1^* = \ldots = \boldsymbol{\beta}_T^* .$$

Finally, we note that equality can be tested for count data with an assumed Poisson distribution.

6.4.2 Ordinal dependent variables

The extension of QGPML to correlated nominal and ordered categorical dependent variables has been discussed, by Miller et al. (1993) and Lipsitz et al. (1994b), respectively. Here, we consider the case of correlated ordinal dependent variables. Let the ordered categorical response of all individuals be coded as $1, \ldots, C$. $C-1$ dummy variables are required to specify all categories. Therefore, the response y_{it} is extended to a response vector $\boldsymbol{y}_{it}$ of length $C-1$, where

$$y_{itc} = \begin{cases} 1, & \text{if } y_{it} \leq c\,, \\ 0, & \text{otherwise}\,. \end{cases}$$

As a result, the response vector $\boldsymbol{y}_i$ of cluster i is of length $T \cdot (C-1)$, and the matrix of independent variables has to be increased accordingly. Specifically, each row $\boldsymbol{x}_{it}'$ of $\boldsymbol{X}_i$ has to be repeated $C-1$ times. Finally, the resulting matrix of independent variables is increased by $C-2$ columns of dummy variables. These are required to model the threshold ϑ of the different categories.

With these specifications, estimation can proceed as before. However, the investigator should check whether the threshold values θ are increasing.

For illustration, we consider the following example.

Example 6.18. We assume that y_{it} has four categories so that three thresholds θ are required. Subsequently, two additional dummy variables are needed. The resulting matrix of independent variables is given by

$$\boldsymbol{X}_i' = \begin{pmatrix} 1 & 0 & 0 & 1 & 0 & 0 & \dots & 1 & 0 & 0 \\ 0 & 1 & 0 & 0 & 1 & 0 & \dots & 0 & 1 & 0 \\ \boldsymbol{x}_{i1} & \boldsymbol{x}_{i1} & \boldsymbol{x}_{i1} & \boldsymbol{x}_{i2} & \boldsymbol{x}_{i2} & \boldsymbol{x}_{i2} & \dots & \boldsymbol{x}_{iT} & \boldsymbol{x}_{iT} & \boldsymbol{x}_{iT} \end{pmatrix},$$

where $\boldsymbol{x}_{it}$ is the vector of independent variables of observation t at cluster i.

The working correlation matrix should make use of the covariance structure of the ordered categorical responses. If the covariance structure of the multinomial distribution is ignored, one can use any working correlation matrix of Sect. 6.3. If the multinomial structure of the data is taken into account, a reasonable working correlation is

$$\boldsymbol{R}_i(\boldsymbol{\alpha}) = \begin{cases} 1 & \text{if } t = t', c = c'\,, \\ \dfrac{-\mu_{itc}\mu_{itc'}}{\sqrt{\mu_{itc}(1-\mu_{itc})\mu_{itc'}(1-\mu_{itc'})}} & \text{if } t = t', c \neq c'\,, \\ \mathbb{C}\mathrm{orr}(y_{itc}, y_{it'c'}) & \text{if } t \neq t', c, c' \text{ arbitrary}\,. \end{cases}$$

A simplification of this working correlation structure is obtained by assuming that the correlation between different time points of an individual equals 0 so that $\mathbb{C}\mathrm{orr}(y_{itc}, y_{it'c'}) = 0$, and the working correlation matrix of a cluster is block diagonal.

Chapter 7
Pseudo maximum likelihood estimation based on the quadratic exponential family

In the last two chapters, we considered approaches for estimating the mean structure, and the association structure was nuisance. Specifically, in Chapt. 5 the covariance matrix was considered fixed, while it was estimated in a two-step procedure in the previous chapter. In many applications, the association structure is, however, of primary interest. Here, we therefore extend the pseudo maximum likelihood (PML) approach to the simultaneous consistent estimation of the mean and the association structure. Because the first two moments are of primary interest, the approach is termed the PML2 method.

This chapter is organized as follows. We define the PML2 estimator in Sect. 7.1 and derive its asymptotic properties in Sect. 7.2. Illustrations of PML2 estimation are provided in Sect. 7.3. Here, four different examples are provided. In the first three examples, we use the second centered moments as the measure of association. Specifically, we first consider the generalized estimating equations of order 2 (GEE2) with an assumed normal distribution. Second, we deal with the special case of dichotomous dependent variables or count data. Third, we use a general quadratic exponential family. Finally, we derive the GEE2 for dichotomous dependent variables using the second ordinary moments as the measure of association. All of these GEE2 have the disadvantage that the set of estimating equations for the mean structure and the association needs to be solved simultaneously. Therefore, a simplification of the GEE2 is desirable, allowing one to separately solve the two estimating equations. The simplification based on the second ordinary moments as the measure of association is termed alternating logistic regression (ALR). The disadvantage of these GEE2 is that they cannot be derived from PML2 estimation although they are a straightforward simplification of the fourth example. However, they can be derived using the generalized method of moments (GMM); see the next chapter. We finally note that GEE2 using the second standardized moments as the measure of association have been proposed in the literature, and they cannot be derived from the PML2 approach but GMM estimation. They are therefore considered in the next chapter.

7.1 Definition

Consider a sample of n independently distributed T-dimensional random vectors $\boldsymbol{y}_i$, and $\boldsymbol{X}_i$ is the $T \times p$ matrix of stochastic and/or fixed explanatory variables of subject i. The true but unknown density (or probability mass function for discrete random vectors) of $\boldsymbol{y}_i$ given $\boldsymbol{X}_i$ is denoted by $f^\star(\boldsymbol{y}_i|\boldsymbol{X}_i)$ with conditional expectation $\boldsymbol{\mu}_i = \mathbb{E}_{f^\star}(\boldsymbol{y}_i|\boldsymbol{X}_i||\boldsymbol{\beta}_0)$ and conditional covariance matrix $\mathbb{V}\mathrm{ar}_{f^\star}(\boldsymbol{y}_i|\boldsymbol{X}_i||\boldsymbol{\beta}_0, \boldsymbol{\alpha}_0) = \boldsymbol{\Omega}_i^\star$.

The true density $f^\star$ may differ from the assumed density f. The assumed conditional pseudo density f of $\boldsymbol{y}_i$ given $\boldsymbol{X}_i$ is parameterized in the $p \times 1$ parameter vector $\boldsymbol{\beta}$ and the $q \times 1$ parameter vector $\boldsymbol{\alpha}$. It is assumed that both the mean structure and the covariance matrix are correctly specified so that $\boldsymbol{\mu}_i = \mathbb{E}_{f^\star}(\boldsymbol{y}_i|\boldsymbol{X}_i) = \mathbb{E}_f(\boldsymbol{y}_i|\boldsymbol{X}_i||\boldsymbol{\beta}_0)$ and $\boldsymbol{\Omega}_i^\star = \mathbb{V}\mathrm{ar}_{f^\star}(\boldsymbol{y}_i|\boldsymbol{X}_i||\boldsymbol{\beta}_0, \boldsymbol{\alpha}_0) = \mathbb{V}\mathrm{ar}_f(\boldsymbol{y}_i|\boldsymbol{X}_i||\boldsymbol{\beta}_0, \boldsymbol{\alpha}_0) = \boldsymbol{\Sigma}_i$.

Estimation is based on a pseudo density from the quadratic exponential family, and the assumed density of cluster i is

$$
\begin{aligned}
f(\boldsymbol{y}_i|\boldsymbol{X}_i||\boldsymbol{\vartheta}_i, \boldsymbol{\lambda}_i) &= \exp\Big(\boldsymbol{\vartheta}_i'\boldsymbol{y}_i - d_i(\boldsymbol{\vartheta}_i, \boldsymbol{\lambda}_i) + b_i(\boldsymbol{y}_i) + \boldsymbol{\lambda}_i'\boldsymbol{w}_i\Big) \\
= f(\boldsymbol{y}_i|\boldsymbol{X}_i||\boldsymbol{\mu}_i, \boldsymbol{\Sigma}_i) &= \exp\Big(\boldsymbol{c}(\boldsymbol{\mu}_i, \boldsymbol{\Sigma}_i)'\boldsymbol{y}_i - d\Big(\boldsymbol{c}(\boldsymbol{\mu}_i, \boldsymbol{\Sigma}_i), \boldsymbol{j}(\boldsymbol{\mu}_i, \boldsymbol{\Sigma}_i)\Big) \\
&\qquad + b(\boldsymbol{y}_i) + \boldsymbol{j}(\boldsymbol{\mu}_i, \boldsymbol{\Sigma}_i)'\boldsymbol{w}_i\Big)\,.
\end{aligned}
$$

Second-order identifiability guarantees the existence of a global maximum at $\boldsymbol{\xi}_0 = (\boldsymbol{\beta}_0', \boldsymbol{\alpha}_0')'$. This means that $\boldsymbol{\mu}(\boldsymbol{X}_i, \boldsymbol{\beta}_0) = \boldsymbol{\mu}(\boldsymbol{X}_i, \boldsymbol{\beta}_1) \Rightarrow \boldsymbol{\beta}_0 = \boldsymbol{\beta}_1$ and $\boldsymbol{\Sigma}(\boldsymbol{X}_i, \boldsymbol{\beta}_0, \boldsymbol{\alpha}_0) = \boldsymbol{\Sigma}(\boldsymbol{X}_i, \boldsymbol{\beta}_0, \boldsymbol{\alpha}_1) \Rightarrow \boldsymbol{\alpha}_0 = \boldsymbol{\alpha}_1$ are required to hold. As a result, the matrix of second derivatives of the pseudo likelihood function is positive definite in a neighborhood of $\boldsymbol{\alpha}_0$.

In summary, PML2 estimation is based on the stochastic model

$$
\begin{aligned}
\boldsymbol{y}_i &= \boldsymbol{\mu}(\boldsymbol{X}_i, \boldsymbol{\beta}_0) + \boldsymbol{\varepsilon}_i \quad \text{with} \quad \mathbb{E}_f(\boldsymbol{\varepsilon}_i|\boldsymbol{X}_i) = \boldsymbol{0} \\
\boldsymbol{\mu}(\boldsymbol{X}_i, \boldsymbol{\beta}_0) &= \mathbb{E}_{f^\star}(\boldsymbol{y}_i|\boldsymbol{X}_i) = \mathbb{E}_f(\boldsymbol{y}_i|\boldsymbol{X}_i) \\
\boldsymbol{\Sigma}_i &= \mathbb{V}\mathrm{ar}_{f^\star}(\boldsymbol{y}_i|\boldsymbol{X}_i||\boldsymbol{\beta}_0, \boldsymbol{\alpha}_0) = \mathbb{V}\mathrm{ar}_f(\boldsymbol{y}_i|\boldsymbol{X}_i||\boldsymbol{\beta}_0, \boldsymbol{\alpha}_0)\,,
\end{aligned}
$$

where the assumed conditional density f belongs to the quadratic exponential family.

Definition 7.1 (PML2 estimator). A pseudo maximum likelihood estimator for the mean and the association structure or, briefly, PML2 estimator of $\boldsymbol{\xi} = (\boldsymbol{\beta}', \boldsymbol{\alpha}')'$ is any value $\hat{\boldsymbol{\xi}} = (\hat{\boldsymbol{\beta}}', \hat{\boldsymbol{\alpha}}')'$ maximizing the kernel of the normed pseudo loglikelihood function

$$
l(\boldsymbol{\beta}, \boldsymbol{\alpha}) = \frac{1}{n}\sum_{i=1}^{n} \ln\exp\Big(\boldsymbol{\vartheta}_i'\boldsymbol{y}_i - d_i(\boldsymbol{\vartheta}_i, \boldsymbol{\lambda}_i) + b_i(\boldsymbol{y}_i) + \boldsymbol{\lambda}_i'\boldsymbol{w}_i\Big)\,.
$$

7.2 Asymptotic properties

In this section, the asymptotic properties of PML2 estimators are formulated. The required regularity conditions and detailed proofs can be found in Gourieroux et al. (1984b, pp. 682, 687, 692). To repeat, a fundamental assumption is that the assumed distribution belongs to the linear exponential family.

Theorem 7.2 (Properties of PML2 estimators).

1. *There asymptotically exists a PML2 estimator* $\hat{\boldsymbol{\xi}}$ *for* $\boldsymbol{\xi}_0$.
2. *The PML2 estimator* $\hat{\boldsymbol{\xi}}$ *converges almost surely to the true parameter vector* $\boldsymbol{\xi}_0$.
3. *Using the second ordinary moments, the score vector of* $\boldsymbol{\xi}$ *is given by*

$$\boldsymbol{u}(\boldsymbol{\xi}) = \boldsymbol{u}\begin{pmatrix}\boldsymbol{\beta}\\ \boldsymbol{\alpha}\end{pmatrix} = \sum_{i=1}^{n} \tilde{\boldsymbol{M}}_i' \tilde{\boldsymbol{V}}_i^{-1} \tilde{\boldsymbol{m}}_i \,,$$

where

$$\tilde{\boldsymbol{M}}_i = \begin{pmatrix} \dfrac{\partial \boldsymbol{\mu}_i}{\partial \boldsymbol{\beta}'} & \boldsymbol{0} \\ \dfrac{\partial \boldsymbol{\nu}_i}{\partial \boldsymbol{\beta}'} & \dfrac{\partial \boldsymbol{\nu}_i}{\partial \boldsymbol{\alpha}'} \end{pmatrix}, \tilde{\boldsymbol{V}}_i = \begin{pmatrix} \boldsymbol{\Sigma}_i & \mathbb{C}ov(\boldsymbol{y}_i, \boldsymbol{w}_i) \\ \mathbb{C}ov(\boldsymbol{w}_i, \boldsymbol{y}_i) & \mathbb{V}ar(\boldsymbol{w}_i) \end{pmatrix}, \tilde{\boldsymbol{m}}_i = \begin{pmatrix} \boldsymbol{y}_i - \boldsymbol{\mu}_i \\ \boldsymbol{w}_i - \boldsymbol{\nu}_i \end{pmatrix}$$

with $\nu_{itt'} = \mathbb{E}_f(y_{it}y_{it'}|\boldsymbol{X}_i)$, $\boldsymbol{\nu}_i = (\nu_{i11}, \nu_{i12}, \ldots, \nu_{iTT})'$, *and* $\boldsymbol{w}_i = (y_{i1}^2, y_{i1}y_{i2}, \ldots, y_{i1}y_{iT}, y_{i2}^2, y_{i2}y_{i3}, \ldots, y_{iT}^2)'$. $\tilde{\boldsymbol{V}}_i$ *is the working covariance matrix consisting of the correctly specified second-order moments* $\boldsymbol{\Sigma}_i$ *and possibly misspecified third- and fourth-order moments.*

4. *The PML2 estimator* $\hat{\boldsymbol{\xi}} = (\hat{\boldsymbol{\beta}}'\hat{\boldsymbol{\alpha}})'$ *for* $\boldsymbol{\xi}_0 = (\boldsymbol{\beta}_0', \boldsymbol{\alpha}_0')'$ *using the second ordinary moments is asymptotically normal. More specifically,*

$$\sqrt{n}(\hat{\boldsymbol{\xi}} - \boldsymbol{\xi}_0) \overset{a}{\sim} N\Big(\boldsymbol{0}, \big(\boldsymbol{A}(\boldsymbol{\xi}_0)\big)^{-1} \boldsymbol{B}(\boldsymbol{\xi}_0) \big(\boldsymbol{A}(\boldsymbol{\xi}_0)\big)^{-1}\Big) \,,$$

where

$$\boldsymbol{A}(\boldsymbol{\xi}) = \mathbb{E}^{\boldsymbol{X}}\big(\mathbb{E}^{\boldsymbol{y}}_{f^\star} - \boldsymbol{W}_i(\boldsymbol{\xi})\big) = \mathbb{E}^{\boldsymbol{X}}\big(\tilde{\boldsymbol{M}}_i' \tilde{\boldsymbol{V}}_i^{-1} \tilde{\boldsymbol{M}}_i\big)$$

and

$$\boldsymbol{B}(\boldsymbol{\xi}_0) = \mathbb{E}^{\boldsymbol{X}}\Big(\mathbb{E}^{\boldsymbol{y}}_{f^\star}\big(\boldsymbol{u}_i(\boldsymbol{\xi})\boldsymbol{u}_i(\boldsymbol{\xi})'\big)\Big) = \mathbb{E}^{\boldsymbol{X}}\big(\tilde{\boldsymbol{M}}_i' \tilde{\boldsymbol{V}}_i^{-1} \tilde{\boldsymbol{\Gamma}}_i \tilde{\boldsymbol{V}}_i^{-1} \tilde{\boldsymbol{M}}_i\big)$$

are the Fisher information matrix of $\boldsymbol{\xi}$ *and the outer product gradient (OPG), respectively.* $\tilde{\boldsymbol{\Gamma}}$ *denotes the covariance matrix* $\mathbb{V}ar_f\big((\boldsymbol{y}_i', \boldsymbol{w}_i')'\big)$ *of* $(\boldsymbol{y}_i', \boldsymbol{w}_i')'$ *under the true distribution* f.

5. *Strongly consistent estimators of* $\boldsymbol{A}(\boldsymbol{\beta}_0, \boldsymbol{\alpha}_0)$ *and* $\boldsymbol{B}(\boldsymbol{\beta}_0, \boldsymbol{\alpha}_0)$ *using the second ordinary moments are given by*

$$\hat{\boldsymbol{A}}(\hat{\boldsymbol{\xi}}) = \hat{\boldsymbol{A}}\begin{pmatrix}\hat{\boldsymbol{\beta}}\\ \hat{\boldsymbol{\alpha}}\end{pmatrix} = \frac{1}{n}\sum_{i=1}^{n}\left(\hat{\tilde{\boldsymbol{M}}}_i'\hat{\tilde{\boldsymbol{V}}}_i^{-1}\hat{\tilde{\boldsymbol{M}}}_i\right)$$

and

$$\hat{\boldsymbol{B}}(\hat{\boldsymbol{\xi}}) = \hat{\boldsymbol{B}}\begin{pmatrix}\hat{\boldsymbol{\beta}}\\ \hat{\boldsymbol{\alpha}}\end{pmatrix} = \frac{1}{n}\sum_{i=1}^{n}\left(\hat{\tilde{\boldsymbol{M}}}_i'\hat{\tilde{\boldsymbol{V}}}_i^{-1}\hat{\tilde{\boldsymbol{\Gamma}}}_i\hat{\tilde{\boldsymbol{V}}}_i^{-1}\hat{\tilde{\boldsymbol{M}}}_i\right),$$

where $\hat{\tilde{\boldsymbol{M}}}_i$ *is the estimated matrix of first derivatives of the mean and the association structure with respect to the parameter vector,* $\hat{\tilde{\boldsymbol{V}}}_i$ *is the estimator of the covariance matrix of the assumed distribution, and* $\tilde{\boldsymbol{\Gamma}}_i$ *is replaced by* $\hat{\tilde{\boldsymbol{\Gamma}}}_i = \hat{\tilde{\boldsymbol{m}}}_i\hat{\tilde{\boldsymbol{m}}}_i' = \binom{\boldsymbol{y}_i-\hat{\boldsymbol{\mu}}_i}{\boldsymbol{w}_i-\hat{\boldsymbol{\nu}}_i}\binom{\boldsymbol{y}_i-\hat{\boldsymbol{\mu}}_i}{\boldsymbol{w}_i-\hat{\boldsymbol{\nu}}_i}'$ *with* $\hat{\boldsymbol{\nu}}_i = \boldsymbol{\nu}(\boldsymbol{X}_i, \hat{\boldsymbol{\beta}}, \hat{\boldsymbol{\alpha}})$.

6. *Using the second central moments, the score vector for* $\boldsymbol{\xi}$ *is given by*

$$\boldsymbol{u}(\boldsymbol{\xi}) = \boldsymbol{u}\begin{pmatrix}\boldsymbol{\beta}\\ \boldsymbol{\alpha}\end{pmatrix} = \sum_{i=1}^{n}\boldsymbol{M}_i'\boldsymbol{V}_i^{-1}\boldsymbol{m}_i\,,$$

where

$$\boldsymbol{M}_i = \begin{pmatrix}\dfrac{\partial\boldsymbol{\mu}_i}{\partial\boldsymbol{\beta}'} & \boldsymbol{0}\\ \dfrac{\partial\boldsymbol{\sigma}_i}{\partial\boldsymbol{\beta}'} & \dfrac{\partial\boldsymbol{\sigma}_i}{\partial\boldsymbol{\alpha}'}\end{pmatrix}, \boldsymbol{V}_i = \begin{pmatrix}\boldsymbol{\Sigma}_i & \mathbb{C}ov(\boldsymbol{y}_i, \boldsymbol{s}_i)\\ \mathbb{C}ov(\boldsymbol{s}_i, \boldsymbol{y}_i) & \mathbb{V}ar(\boldsymbol{s}_i)\end{pmatrix}, \boldsymbol{m}_i = \begin{pmatrix}\boldsymbol{y}_i - \boldsymbol{\mu}_i\\ \boldsymbol{s}_i - \boldsymbol{\sigma}_i\end{pmatrix}$$

with $\boldsymbol{s}_i = (s_{i11}, s_{i12}, \ldots, s_{iTT})'$, $\boldsymbol{\sigma}_i = (\sigma_{i11}, \sigma_{i12}, \ldots, \sigma_{iTT})'$, $s_{itt'} = (y_{it} - \mu_{it})(y_{it'} - \mu_{it'})$, *and* $\sigma_{itt'} = \nu_{itt'} - \mu_{it}\mu_{it'}$. $\boldsymbol{V}_i$ *is the working covariance matrix consisting of correctly specified second-order moments* $\boldsymbol{\Sigma}_i$ *and possibly misspecified third- and fourth-order moments.*

7. *The PML2 estimator* $\hat{\boldsymbol{\xi}} = (\hat{\boldsymbol{\beta}}'\hat{\boldsymbol{\alpha}})'$ *for* $\boldsymbol{\xi}_0 = (\boldsymbol{\beta}_0', \boldsymbol{\alpha}_0')'$ *using the second central moments is asymptotically normal. More specifically,*

$$\sqrt{n}(\hat{\boldsymbol{\xi}} - \boldsymbol{\xi}_0) \overset{a}{\sim} N\Big(\boldsymbol{0}, \big(\boldsymbol{A}(\boldsymbol{\beta}_0, \boldsymbol{\alpha}_0)\big)^{-1}\boldsymbol{B}(\boldsymbol{\beta}_0, \boldsymbol{\alpha}_0)\big(\boldsymbol{A}(\boldsymbol{\beta}_0, \boldsymbol{\alpha}_0)\big)^{-1}\Big),$$

where

$$\boldsymbol{A}(\boldsymbol{\xi}) = \mathbb{E}^{\boldsymbol{X}}\big(\mathbb{E}_{f^\star}^{\boldsymbol{y}} - \boldsymbol{W}_i(\boldsymbol{\xi})\big) = \mathbb{E}^{\boldsymbol{X}}\big(\boldsymbol{M}_i'\boldsymbol{V}_i^{-1}\boldsymbol{M}_i\big)$$

and

$$\boldsymbol{B}(\boldsymbol{\xi}_0) = \mathbb{E}^{\boldsymbol{X}}\Big(\mathbb{E}_{f^\star}^{\boldsymbol{y}}\big(\boldsymbol{u}_i(\boldsymbol{\xi})\boldsymbol{u}_i(\boldsymbol{\xi})'\big)\Big) = \mathbb{E}^{\boldsymbol{X}}\big(\boldsymbol{M}_i'\boldsymbol{V}_i^{-1}\boldsymbol{\Gamma}_i\boldsymbol{V}_i^{-1}\boldsymbol{M}_i\big)$$

are the Fisher information matrix of $\boldsymbol{\xi}$ *and the OPG, respectively, and* $\boldsymbol{\Gamma}$ *denotes the covariance matrix* $\mathbb{V}\mathrm{ar}_f\big((\boldsymbol{y}_i', \boldsymbol{s}_i')'\big)$ *of* $(\boldsymbol{y}_i', \boldsymbol{s}_i')'$ *under the true distribution* f.

8. *Strongly consistent estimators of* $\boldsymbol{A}(\boldsymbol{\beta}_0, \boldsymbol{\alpha}_0)$ *and* $\boldsymbol{B}(\boldsymbol{\beta}_0, \boldsymbol{\alpha}_0)$ *using the second central moments are given by*

$$\hat{\boldsymbol{A}}(\hat{\boldsymbol{\xi}}) = \hat{\boldsymbol{A}}\begin{pmatrix}\hat{\boldsymbol{\beta}}\\ \hat{\boldsymbol{\alpha}}\end{pmatrix} = \frac{1}{n}\sum_{i=1}^{n}\Big(\hat{\boldsymbol{M}}_i'\hat{\boldsymbol{V}}_i^{-1}\hat{\boldsymbol{M}}_i\Big)$$

and

$$\hat{\boldsymbol{B}}(\hat{\boldsymbol{\xi}}) = \hat{\boldsymbol{B}}\begin{pmatrix}\hat{\boldsymbol{\beta}}\\ \hat{\boldsymbol{\alpha}}\end{pmatrix} = \frac{1}{n}\sum_{i=1}^{n}\Big(\hat{\boldsymbol{M}}_i'\hat{\boldsymbol{V}}_i^{-1}\hat{\boldsymbol{\Gamma}}_i\hat{\boldsymbol{V}}_i^{-1}\hat{\boldsymbol{M}}_i\Big)\,,$$

where $\hat{\boldsymbol{M}}_i$ *is the estimated matrix of first derivatives of the mean and the association structure with respect to the parameter vector,* $\hat{\boldsymbol{V}}_i$ *is the estimator of the covariance matrix of the assumed distribution, and* $\boldsymbol{\Gamma}_i$ *is replaced by* $\hat{\boldsymbol{\Gamma}}_i = \hat{\boldsymbol{m}}_i\hat{\boldsymbol{m}}_i' = \begin{pmatrix}\boldsymbol{y}_i - \hat{\boldsymbol{\mu}}_i\\ \boldsymbol{s}_i - \hat{\boldsymbol{\sigma}}_i\end{pmatrix}\begin{pmatrix}\boldsymbol{y}_i - \hat{\boldsymbol{\mu}}_i\\ \boldsymbol{s}_i - \hat{\boldsymbol{\sigma}}_i\end{pmatrix}'$ *with* $\hat{\boldsymbol{\sigma}}_i = \boldsymbol{\sigma}(\boldsymbol{X}_i, \hat{\boldsymbol{\beta}}, \hat{\boldsymbol{\alpha}})$.

9. *Necessary for the strong consistency of a PML2 estimator associated with a family of assumed distributions* $f(\boldsymbol{y}_i|\boldsymbol{X}_i||\boldsymbol{\beta}, \boldsymbol{\alpha})$ *for any parameter space, parameter vector* $\boldsymbol{\xi} = (\boldsymbol{\beta}', \boldsymbol{\alpha}')'$, *mean structure, association structure, and true distribution* $f^\star$ *is that the assumed distribution belongs to the quadratic exponential family.*

Remark 7.3.

- One assumption in the estimating equations is $\partial\boldsymbol{\mu}_i/\partial\boldsymbol{\alpha}' = \boldsymbol{0}$. In fact, the mean structure $\boldsymbol{\mu}_i$ should depend only on the mean structure parameter $\boldsymbol{\beta}$, but it should be independent of the parameter $\boldsymbol{\alpha}$ characterizing the association structure. However, the association structures $\boldsymbol{\nu}_i$ and $\boldsymbol{\sigma}_i$ may depend on the mean structure parameter. For example, the variance v_{it} in GLM generally is a function of the mean μ_{it}.
- The estimating equations generally need to be solved jointly, i.e., in a one-step procedure. A separation in one set of estimating equations for $\boldsymbol{\beta}$ and another one for $\boldsymbol{\alpha}$ is only possible under specific assumptions, which will be discussed below.
- If investigators assume the independence of the association structure and the mean structure parameter, i.e., if they assume, e.g., $\partial\boldsymbol{\sigma}_i/\partial\boldsymbol{\beta}' = \boldsymbol{0}$, the matrix of first derivatives $\boldsymbol{M}_i$ is block diagonal. As a result, the estimating equations reduce to

$$\boldsymbol{0} = \hat{\boldsymbol{u}}\begin{pmatrix}\boldsymbol{\beta}\\ \boldsymbol{\alpha}\end{pmatrix} = \sum_{i=1}^{n}\begin{pmatrix}\dfrac{\partial\hat{\boldsymbol{\mu}}_i'}{\partial\boldsymbol{\beta}} & \boldsymbol{0}\\ \boldsymbol{0} & \dfrac{\partial\hat{\boldsymbol{\sigma}}_i'}{\partial\boldsymbol{\alpha}}\end{pmatrix}\begin{pmatrix}\hat{\boldsymbol{\Sigma}}_i & \widehat{\mathbb{C}\mathrm{ov}}(\boldsymbol{y}_i, \boldsymbol{s}_i)\\ \widehat{\mathbb{C}\mathrm{ov}}(\boldsymbol{s}_i, \boldsymbol{y}_i) & \widehat{\mathbb{V}\mathrm{ar}}(\boldsymbol{s}_i)\end{pmatrix}^{-1}\begin{pmatrix}\boldsymbol{y}_i - \hat{\boldsymbol{\mu}}_i\\ \boldsymbol{s}_i - \hat{\boldsymbol{\sigma}}_i\end{pmatrix}.$$

These estimating equations can be separated, and one obtains for the first set of estimating equations

$$\mathbf{0} = \sum_{i=1}^{n} \left(\frac{\partial \hat{\boldsymbol{\mu}}_i'}{\partial \boldsymbol{\beta}} \hat{\boldsymbol{\Sigma}}_i^{-1} (\boldsymbol{y}_i - \hat{\boldsymbol{\mu}}_i) + \frac{\partial \hat{\boldsymbol{\mu}}_i'}{\partial \boldsymbol{\beta}} \widehat{\mathbb{C}\mathrm{ov}}(\boldsymbol{y}_i, \boldsymbol{s}_i)^{-1} (\boldsymbol{s}_i - \hat{\boldsymbol{\sigma}}_i) \right) .$$

Thus, the parameters need to be orthogonal for consistent estimation of $\boldsymbol{\beta}$. This means that the covariance matrix $\mathbb{V}\mathrm{ar}\big((\boldsymbol{y}_i', \boldsymbol{s}_i')'\big)$ needs to be block diagonal. Using the orthogonality condition, the estimating equations can be rewritten as

$$\mathbf{0} = \sum_{i=1}^{n} \begin{pmatrix} \dfrac{\partial \hat{\boldsymbol{\mu}}_i'}{\partial \boldsymbol{\beta}} & \mathbf{0} \\ \mathbf{0} & \dfrac{\partial \hat{\boldsymbol{\sigma}}_i'}{\partial \boldsymbol{\alpha}} \end{pmatrix} \begin{pmatrix} \widehat{\mathbb{V}\mathrm{ar}}(\boldsymbol{y}_i) & \mathbf{0} \\ \mathbf{0} & \widehat{\mathbb{V}\mathrm{ar}}(\boldsymbol{s}_i) \end{pmatrix}^{-1} \begin{pmatrix} \boldsymbol{y}_i - \hat{\boldsymbol{\mu}}_i \\ \boldsymbol{s}_i - \hat{\boldsymbol{\sigma}}_i \end{pmatrix} .$$

These estimating equations are generally termed "ad hoc estimating equations" in the literature.
If parameters are indeed orthogonal, these estimating equations yield consistent and jointly asymptotically normally distributed parameter estimators $\hat{\boldsymbol{\beta}}$ and $\hat{\boldsymbol{\alpha}}$. If these estimating equations are used when $\partial \boldsymbol{\sigma}_i'/\partial \boldsymbol{\beta} = \mathbf{0}$ does not hold, $\hat{\boldsymbol{\beta}}$ will remain consistent for $\boldsymbol{\beta}$, but $\hat{\boldsymbol{\alpha}}$ will not be a consistent estimator for $\boldsymbol{\alpha}$, in general.
In most applications, this will be the case since $\boldsymbol{\alpha}$ is not defined via $\boldsymbol{\sigma}_i$ but internally using a "working correlation matrix" (cf. Section 5.1) and a transformation from the second standardized to the second central moments.

Proof. The proof follows the lines as the proof of the asymptotic properties for PML1 estimation (Theorem 5.2) because the quadratic exponential family can be embedded in the framework of the linear exponential family. In detail,
1.: Existence: See White (1981, Theorem 2.1).
2.: Consistency: The proof is carried out analogously to the proof of the consistency of Theorem 5.2 for which a detailed proof has been given, e.g., by Gourieroux et al. (1984b, Theorem 4). To prove the strong consistency, we have to show that the expected value of the kernel of the loglikelihood has a unique maximum at $\boldsymbol{\xi}_0 = (\boldsymbol{\beta}_0', \boldsymbol{\alpha}_0')'$. This is true because

$$\begin{aligned} \mathbb{E}^{\boldsymbol{X}} \mathbb{E}^{\boldsymbol{y}}_{\boldsymbol{\xi}_0}\big(l(\boldsymbol{\xi})\big) &= \boldsymbol{c}(\boldsymbol{\mu}_{i0}, \boldsymbol{\Sigma}_{i0})' \mathbb{E}^{\boldsymbol{X}} \mathbb{E}^{\boldsymbol{y}}_{\boldsymbol{\xi}_0}(\boldsymbol{y}_i) + a(\boldsymbol{\mu}_{i0}, \boldsymbol{\Sigma}_{i0}) + \boldsymbol{j}(\boldsymbol{\mu}_{i0}, \boldsymbol{\Sigma}_{i0})' \mathbb{E}^{\boldsymbol{X}} \mathbb{E}^{\boldsymbol{y}}_{\boldsymbol{\xi}_0}(\boldsymbol{w}_i) \\ &= \boldsymbol{c}(\boldsymbol{\mu}_{i0}, \boldsymbol{\Sigma}_{i0})' \boldsymbol{\mu}_{i0} + a(\boldsymbol{\mu}_{i0}, \boldsymbol{\Sigma}_{i0}) + \boldsymbol{j}(\boldsymbol{\mu}_{i0}, \boldsymbol{\Sigma}_{i0})' \boldsymbol{\nu}_{i0} , \end{aligned}$$

where $\boldsymbol{\mu}_{i0} = \boldsymbol{\mu}(\boldsymbol{X}_i, \boldsymbol{\beta}_0)$ and $\boldsymbol{\nu}_{i0}$ is defined analogously. The result now follows from Property 2.3 and the second-order identifiability of $\boldsymbol{\beta}$ and $\boldsymbol{\alpha}$.
3.: Score equations using the second ordinary moments: Differentiation of the kernel of the individual pseudo loglikelihood function based on the quadratic exponential family $l_i(\boldsymbol{\beta}, \boldsymbol{\alpha}) = \boldsymbol{\vartheta}_i' \boldsymbol{y}_i - d(\boldsymbol{\vartheta}_i, \boldsymbol{\lambda}_i) + \boldsymbol{\lambda}_i' \boldsymbol{w}_i$ with respect to $\boldsymbol{\beta}$ and

$\boldsymbol{\alpha}$ yields

$$\boldsymbol{u}(\boldsymbol{\xi}) = \boldsymbol{u}\begin{pmatrix}\boldsymbol{\beta}\\ \boldsymbol{\alpha}\end{pmatrix} = \frac{\partial l_i}{\partial \begin{pmatrix}\boldsymbol{\beta}\\ \boldsymbol{\alpha}\end{pmatrix}} = \frac{\partial(\boldsymbol{\mu}_i', \boldsymbol{\nu}_i')}{\partial \begin{pmatrix}\boldsymbol{\beta}\\ \boldsymbol{\alpha}\end{pmatrix}} \cdot \frac{\partial(\boldsymbol{\vartheta}_i', \boldsymbol{\lambda}_i')}{\partial \begin{pmatrix}\boldsymbol{\mu}\\ \boldsymbol{\nu}\end{pmatrix}} \cdot \frac{\partial l_i}{\partial \begin{pmatrix}\boldsymbol{\vartheta}_i\\ \boldsymbol{\lambda}_i\end{pmatrix}}$$

$$= \begin{pmatrix} \dfrac{\partial \boldsymbol{\mu}_i}{\partial \boldsymbol{\beta}'} & \dfrac{\partial \boldsymbol{\mu}_i}{\partial \boldsymbol{\alpha}'} \\[2ex] \dfrac{\partial \boldsymbol{\nu}_i}{\partial \boldsymbol{\beta}'} & \dfrac{\partial \boldsymbol{\nu}_i}{\partial \boldsymbol{\alpha}'} \end{pmatrix}' \begin{pmatrix} \dfrac{\partial \boldsymbol{\vartheta}_i}{\partial \boldsymbol{\mu}'} & \dfrac{\partial \boldsymbol{\vartheta}_i}{\partial \boldsymbol{\nu}'} \\[2ex] \dfrac{\partial \boldsymbol{\lambda}_i}{\partial \boldsymbol{\mu}'} & \dfrac{\partial \boldsymbol{\lambda}_i}{\partial \boldsymbol{\nu}'} \end{pmatrix} \begin{pmatrix} \dfrac{\partial l_i}{\partial \boldsymbol{\vartheta}} \\[2ex] \dfrac{\partial l_i}{\partial \boldsymbol{\lambda}} \end{pmatrix} = \tilde{\boldsymbol{M}}_i' \tilde{\boldsymbol{V}}_i^{-1} \tilde{\boldsymbol{m}}_i \,. \tag{7.1}$$

Summation yields the required result since $\partial \boldsymbol{\mu}_i / \partial \boldsymbol{\alpha}' = \mathbf{0}$.
4.: The asymptotic normality can be shown analogously to the asymptotic normality of Theorem 5.2.
5.: Estimation: See White (1982, Theorem 3.2) using the argument that the properties of the linear exponential family can be carried over to the linear exponential family.
6.: Score equations using the second central moments: For a transformation to the second central moments $\nu_{itt'} = \sigma_{itt'} + \mu_{it}\mu_{it'}$, a block matrix $\boldsymbol{Q}_i$ is used in order to write $\tilde{\boldsymbol{m}}_i = \boldsymbol{Q}_i \boldsymbol{m}_i$, $\tilde{\boldsymbol{V}}_i = \boldsymbol{Q}_i \boldsymbol{V}_i \boldsymbol{Q}_i'$ and $\tilde{\boldsymbol{M}}_i = \boldsymbol{Q}_i \boldsymbol{M}_i$. The block matrix $\boldsymbol{Q}_i$ yielding the desired result is given by

$$\boldsymbol{Q}_i = \begin{pmatrix} \boldsymbol{I}_T & \mathbf{0} \\ \boldsymbol{L}_i & \boldsymbol{I}_{T(T+1)/2} \end{pmatrix},$$

where $\boldsymbol{I}_T$ is the $T \times T$ identity matrix and $\boldsymbol{L}_i' = (\boldsymbol{L}_{i1}', \ldots, \boldsymbol{L}_{iT}')$ is defined by

$$\boldsymbol{L}_{i1} = \begin{pmatrix} 2\mu_{i1} & & & \mathbf{0} \\ \mu_{i2} & \mu_{i1} & & \\ \vdots & & \ddots & \vdots \\ \mu_{iT} & \mathbf{0} & & \mu_{i1} \end{pmatrix}, \quad T \times T,$$

$$\boldsymbol{L}_{i2} = \begin{pmatrix} 0 & 2\mu_{i2} & & & \mathbf{0} \\ 0 & \mu_{i3} & \mu_{i2} & & \\ \vdots & \vdots & & \ddots & \vdots \\ 0 & \mu_{iT} & \mathbf{0} & & \mu_{i2} \end{pmatrix}, \ (T-1) \times T,$$

$$\vdots$$

$$\boldsymbol{L}_{it} = \underbrace{\begin{pmatrix} 0 \cdots 0 & 2\mu_{it} & & & \boldsymbol{0} \\ 0 \quad\; 0 & \mu_{i,t+1} & \mu_{it} & & \\ \vdots \quad\; \vdots & \vdots & & \ddots & \\ 0 \cdots 0 & \mu_{iT} & \boldsymbol{0} & & \mu_{it} \end{pmatrix}}_{t-1}, \; (T-t) \times T \, ,$$

$$\vdots$$

$$\boldsymbol{L}_{i,T-1} = \begin{pmatrix} 0 \cdots 0 & 2\mu_{i,T-1} & 0 \\ 0 \cdots 0 & \mu_{iT} & \mu_{i,T-1} \end{pmatrix}, \qquad (2 \times T)\,,$$

$$\boldsymbol{L}_{iT} = (0, \ldots, 0, 2\mu_{iT})\,, \qquad (1 \times T)\,.$$

As a result, Eq. 7.1 can be rewritten as

$$\boldsymbol{u}_i(\boldsymbol{\beta}', \boldsymbol{\alpha}')' = \boldsymbol{M}_i'\boldsymbol{V}_i^{-1}\boldsymbol{m}_i = \tilde{\boldsymbol{M}}_i'\boldsymbol{L}_i'\boldsymbol{L}_i'^{-1}\tilde{\boldsymbol{V}}_i^{-1}\boldsymbol{L}_i'\boldsymbol{L}_i'^{-1}\tilde{\boldsymbol{m}}_i = \tilde{\boldsymbol{M}}_i'\tilde{\boldsymbol{V}}_i^{-1}\tilde{\boldsymbol{m}}_i\,.$$

The modified Fisher information matrix and the OPG can be obtained as before using the transformation matrix $\boldsymbol{U}_i$.
8.: Estimation: See White (1982, Theorem 3.2).
9.: Necessary condition for strong consistency: See Gourieroux et al. (1984b, Appendix 3). □

In the following, we give a simple example for PML2 estimation. More complex examples, including several GEE2, are provided in the following sections.

Example 7.4 (Difference between two means with common variance). Consider the simple two-sample scenario, where $y_{11}, \ldots, y_{1n_1}$ and $y_{21}, \ldots, y_{2n_2}$ are independently identically distributed, y_{2i} have mean μ_1 and y_{2i} have mean μ_2, and there is a common variance σ^2 (Royall, 1986). The normal distribution is chosen as assumed distribution.

The loglikelihood of all $n_1 + n_2$ observations is given by

$$\begin{aligned} l(\mu_1, \mu_2, \sigma^2) = &- \tfrac{1}{2}(n_1+n_2)\ln\sigma^2 - \frac{1}{2}(n_1+n_2)\ln(2\pi) \\ &- \frac{1}{2}\sum_{i=1}^{n_1}\frac{(y_{1i}-\mu_1)^2}{\sigma^2} - \frac{1}{2}\sum_{i=1}^{n_2}\frac{(y_{2i}-\mu_2)^2}{\sigma^2}\,. \end{aligned}$$

In this example, $\boldsymbol{\beta} = (\mu_1, \mu_2)'$, and $\alpha = \sigma^2$. The regression matrix $\boldsymbol{X}_i$ of individual i is either the vector $(1, 0)'$, if i belongs to the first group, or the vector $(0, 1)'$, if i belongs to the second group of observations. For $g = 1, 2$ denoting the group, first- and second-order derivatives are given by

$$\frac{\partial l}{\partial \mu_g} = \frac{1}{\sigma^2}\sum_{i=1}^{n_g}(y_{gi}-\mu_g)\,, \qquad \frac{\partial^2 l}{\partial \mu_g^2} = -\frac{n_g}{\sigma^2}\,, \qquad \frac{\partial^2 l}{\partial \mu_1 \partial \mu_2} = \frac{\partial^2 l}{\partial \mu_2 \partial \mu_1} = 0\,,$$

$$\frac{\partial l}{\partial \sigma^2} = -\frac{1}{2}\frac{n_1+n_2}{\sigma^2} + \frac{1}{2}\sum_{i=1}^{n_1}\frac{(y_{1i}-\mu_1)^2}{\sigma^4} + \frac{1}{2}\sum_{i=1}^{n_2}\frac{(y_{2i}-\mu_2)^2}{\sigma^4}\,,$$

$$\frac{\partial^2 l}{\partial (\sigma^2)^2} = \frac{1}{2}\frac{n_1+n_2}{\sigma^4} - \sum_{i=1}^{n_1}\frac{(y_{1i}-\mu_1)^2}{\sigma^6} - \sum_{i=1}^{n_2}\frac{(y_{2i}-\mu_2)^2}{\sigma^6}\,,$$

$$\frac{\partial^2 l}{\partial \mu_g \partial \sigma^2} = \frac{\partial^2 l}{\partial \sigma^2 \partial \mu_g} = -\frac{1}{\sigma^4}\sum_{i=1}^{n_g}(y_{gi}-\mu_g)\,.$$

The estimating equations for $\boldsymbol{\beta} = (\mu_1, \mu_2)'$ and the parameter of association $\alpha = \sigma^2$ are given by

$$\begin{pmatrix} \hat{\mu_1} \\ \hat{\mu_2} \\ \hat{\sigma}^2 \end{pmatrix} = \begin{pmatrix} \bar{y_1} \\ \bar{y_2} \\ \dfrac{n_1 \sum_{i=1}^{n_1}(y_{1i}-\hat{\mu}_1)^2 + n_2 \sum_{i=1}^{n_2}(y_{2i}-\hat{\mu}_2)^2}{n_1+n_2} \end{pmatrix} = \begin{pmatrix} \bar{y_1} \\ \bar{y_2} \\ \dfrac{n_1\hat{\sigma}_1^2 + n_2\hat{\sigma}_2^2}{n_1+n_2} \end{pmatrix}.$$

As a result, the expected Fisher information matrix, i.e., the model-based covariance matrix, is diagonal, while the OPG is not diagonal. The Fisher information matrix can be estimated by

$$\hat{\boldsymbol{A}}\begin{pmatrix} \hat{\mu} \\ \hat{\nu} \\ \hat{\sigma}^2 \end{pmatrix} = \operatorname{diag}\left(\frac{n_1}{\hat{\sigma}^2}, \frac{n_2}{\hat{\sigma}^2}, \frac{1}{2}\frac{n_1+n_2}{\hat{\sigma}^4}\right).$$

The robust covariance matrix can be estimated by

$$\hat{\boldsymbol{C}} = \begin{pmatrix} \dfrac{\hat{\sigma}_1^2}{n_1} & 0 & \dfrac{\sum_{i=1}^{n_1}(y_{1i}-\bar{y_1})^3}{n_1(n_1+n_2)} \\ 0 & \dfrac{\hat{\sigma}_2^2}{n_2} & \dfrac{\sum_{i=1}^{n_2}(y_{2i}-\bar{y_1})^3}{n_2(n_1+n_2)} \\ \dfrac{\sum_{i=1}^{n_1}(y_{1i}-\bar{y_1})^3}{n_1(n_1+n_2)} & \dfrac{\sum_{i=1}^{n_2}(y_{2i}-\bar{y_1})^3}{n_2(n_1+n_2)} & [\hat{\boldsymbol{C}}]_{33} \end{pmatrix}$$

with

$$[\hat{\boldsymbol{C}}]_{33} = \frac{1}{(n_1+n_2)^2}\left(\sum_{i=1}^{n_1}\left((y_{1i}-\bar{y_1})^2-\hat{\sigma}^2\right)^2 + \sum_{i=1}^{n_2}\left((y_{2i}-\bar{y_2})^2-\hat{\sigma}^2\right)^2\right).$$

Thus, robust confidence intervals for $\mu_1 \pm \mu_2$ would replace the common maximum likelihood variance estimator $(n_1+n_2)\hat{\sigma}^2/(n_1 n_2)$ by $\hat{\sigma}_1^2/n_1+\hat{\sigma}_2^2/n_2$.

Furthermore, $[\hat{\boldsymbol{C}}]_{33}$ can be used to derive a robust confidence interval for $\hat{\sigma}^2$ that is still valid, if the assumption of normality fails.

7.3 Examples

In this section, four examples for GEE are given, and we start with general GEE2 using an assumed normal distribution and the second centered moments as the measure of association.

7.3.1 Generalized estimating equations 2 with an assumed normal distribution using the second centered moments

For continuous dependent variables, the domain of μ_{it} needs to be the real line. Furthermore, σ_{it}^2 needs to be specified because it possibly is independent of the mean μ_{it}. If the normal distribution is used as assumed distribution for GEE2 estimation, the first two moments specify all higher order moments. Thus, a working covariance or working correlation matrix for third- and fourth-order moments cannot be chosen, but it is fixed after specification of the first two moments. At the same time, the advantage is that they need not be estimated separately.

The assumed distribution of $\boldsymbol{y}_i$ given $\boldsymbol{X}_i$ is the normal distribution with mean from a generalized linear model (GLM), i.e., $\boldsymbol{\mu}_i = \mu_{i1}, \ldots, \mu_{iT})'$, $\mathbb{E}(y_{it}|\boldsymbol{x}_{it}) = \mu_{it} = g(\boldsymbol{x}_{it}'\boldsymbol{\beta}_0)$ for which we write $g(\boldsymbol{X}_i'\boldsymbol{\beta}_0)$ as in the previous chapters. Furthermore, the covariance matrix is $\boldsymbol{\Sigma}_i$, which is a function of $\boldsymbol{\beta}_0$ and $\boldsymbol{\alpha}_0$ so that

$$\boldsymbol{y}_i|\boldsymbol{X}_i \sim \mathrm{N}\big(g(\boldsymbol{X}_i'\boldsymbol{\beta}_0), \boldsymbol{\Sigma}_i(\boldsymbol{X}_i, \boldsymbol{\beta}_0, \boldsymbol{\alpha}_0)\big) .$$

The GEE2 using the second centered moments as the measure of association are given by

$$\boldsymbol{0} = \boldsymbol{u}(\hat{\boldsymbol{\xi}}) = \boldsymbol{u}\begin{pmatrix} \hat{\boldsymbol{\beta}} \\ \hat{\boldsymbol{\alpha}} \end{pmatrix} = \sum_{i=1}^{n} \hat{\boldsymbol{M}}_i' \hat{\boldsymbol{\Sigma}}_i^{-1} \hat{\boldsymbol{m}}_i .$$

Because the normal distribution is chosen as assumed distribution, third-order moments are given by $\mathbb{C}\mathrm{ov}(\boldsymbol{y}_i, \boldsymbol{s}_i) = \boldsymbol{0}$, and fourth-order moments are (Anderson, 1984, p. 49)

$$\begin{aligned}\mathbb{C}\text{ov}(s_{i,tt'}, s_{i,rr'}) &= \mathbb{E}\big((y_{it}-\mu_{it})(y_{it'}-\mu_{it'})(y_{ir}-\mu_{ir})(y_{ir'}-\mu_{ir'})\big)\\ &\quad - \sigma_{i,tt'}\sigma_{i,rr'}\\ &= \sigma_{i,tr}\sigma_{i,t'r'} + \sigma_{i,tr'}\sigma_{i,t'r}\,.\end{aligned}$$

This completes the specification of the working covariance matrix.

We stress that although third- and fourth-order moments may be misspecified, both first- and second-order moments need to be correctly specified for consistent estimation of both $\boldsymbol{\beta}_0$ and $\boldsymbol{\alpha}_0$. We have already assumed the specification of the mean structure through a link function from the GLM. However, the association function, i.e., $\sigma_{itt'} = \mathbb{C}\text{ov}(y_{it}, y_{it'})$, also needs to be specified as a function of $\boldsymbol{\beta}$ and $\boldsymbol{\alpha}$ and the matrix of explanatory variables $\boldsymbol{X}_i$ of subject i.

To model $\sigma_{itt'}$, Prentice and Zhao (1991) used the transformation

$$\sigma_{itt'} = \frac{\mathbb{C}\text{orr}(y_{it}, y_{it'})}{\sqrt{\sigma_{it}^2\sigma_{it'}^2}}\,, = \frac{\boldsymbol{k}(\boldsymbol{x}_{it}, \boldsymbol{x}_{it'})^T\boldsymbol{\alpha}_{tt'}}{\sigma_{it}\sigma_{it'}}$$

and they established a functional relationship between $\sigma_{itt'}$ and $\boldsymbol{\alpha}$ using the correlation coefficient:

$$\mathbb{C}\text{orr}(y_{it}, y_{it'}) = \boldsymbol{k}(\boldsymbol{x}_{it}, \boldsymbol{x}_{it'})'\boldsymbol{\alpha}_{tt'}\,.$$

Here, $\boldsymbol{k}(\boldsymbol{x}_{it}, \boldsymbol{x}_{it'})$ is a pre-specified function of $\boldsymbol{x}_{it}$ and $\boldsymbol{x}_{it'}$, which is a linear function of $\boldsymbol{\alpha}$. Because the correlation coefficient is restricted to the interval $[-1; 1]$, Fisher's z transformation is generally used as an association link function:

$$\mathbb{C}\text{orr}(y_{it}, y_{it'}) = \frac{\exp\big(\boldsymbol{k}(\boldsymbol{x}_{it}, \boldsymbol{x}_{it'})'\boldsymbol{\alpha})\big) - 1}{\exp\big(\boldsymbol{k}(\boldsymbol{x}_{it}, \boldsymbol{x}_{it'})'\boldsymbol{\alpha})\big) + 1}\,.$$

Finally, the variance σ_{it}^2 needs to be specified. For most dependent variables, the variance function from a GLM can be used. For continuous data, the log link to guarantee positivity,

$$\sigma_{it}^2 = \exp(\boldsymbol{x}_{it}'\boldsymbol{\alpha})\,, \tag{7.2}$$

is often used. This completes the model specification. This variance is assumed to be independent of $\boldsymbol{\beta}$. In the literature, the association parameter for the variance function (7.2) is often denoted by ϕ, and a set of three estimating equations, i.e., for the mean structure, the variance function, and the correlation coefficient, is constructed (see, e.g., Yan and Fine, 2004). By considering the GEE2 as a special case of PML2 estimation, this distinction is not required.

The asymptotic normality of the resulting estimator $\hat{\boldsymbol{\xi}} = (\hat{\boldsymbol{\beta}}', \hat{\boldsymbol{\alpha}}')'$ follows from Theorem 7.2. Estimation based on the normal distribution chosen as assumed distribution is reasonable, if the true distribution is close to the normal distribution. Furthermore, implementation in a computer package is convenient. Computation time needed for estimation is low, because moments of order three are 0, and fourth-order moments are calculated from lower order moments. This is especially important for applications with categorical dependent variables. For example, if $\boldsymbol{y}_i$ can take three different values and if $T = 4$, estimation of third- and fourth-order moments requires summation over $3^3 = 27$ and $3^4 = 81$ terms (Prentice and Zhao, 1991, p. 830). Estimation of these moments has to be repeated during the iterations. Therefore, it is common to express third- and fourth-order moments as functions of first- and second-order moments.

7.3.2 Generalized estimating equations 2 for binary data or count data with an assumed normal distribution using the second centered moments

In this short section, we specifically consider the case of dichotomous dependent variables. To this end, we assume that $\boldsymbol{y}_i$ is a vector of binary random variables. The mean structure is assumed to be defined by $\mathbb{E}(y_{it}|\boldsymbol{x}_{it}) = \mu_{it} = F(\boldsymbol{x}_{it}'\boldsymbol{\beta}_0)$ for some cumulative distribution function F. The variance function from the binomial distribution is chosen so that $\sigma_{it}^2 = \mu_{it}(1 - \mu_{it})$. Finally, we use $\sigma_{itt'} = \mathbb{C}\text{orr}(y_{it}, y_{it'})/(\sigma_{it}\sigma_{it'})$ to model the relationship between the covariance $\sigma_{itt'}$ and $\boldsymbol{\alpha}$. As in the previous section, we choose Fisher's z as an association link:

$$\mathbb{C}\text{orr}(y_{it}, y_{it'}) = \frac{\exp\big(\boldsymbol{k}(\boldsymbol{x}_{it}, \boldsymbol{x}_{it'})'\boldsymbol{\alpha})\big) - 1}{\exp\big(\boldsymbol{k}(\boldsymbol{x}_{it}, \boldsymbol{x}_{it'})'\boldsymbol{\alpha})\big) + 1}\,.$$

In some cases, estimation using the normal distribution as assumed distribution is not very efficient because higher order moments are neglected in the quadratic exponential family. Nevertheless, the GEE2 with assumed normal distribution may be applied, and it yields consistent parameter estimates of both the mean and the association structure if first- and second-order moments are correctly specified.

The use of the assumed normal distribution for dichotomous dependent variables allows the following simplification. PML2 estimation includes a function for the conditional variances $\sigma_{it}^2 = \mathbb{V}\text{ar}(y_{it}|\boldsymbol{X}_i)$ because these need not be completely specified by the conditional mean $\mathbb{E}(y_{it}|\boldsymbol{X}_i)$. With the assumption of the binomial distribution, the conditional variance is a function of the mean, and the variances can be omitted from estimation.

The number of parameters of the covariance matrix to be estimated is reduced to $\frac{T(T-1)}{2}$ from $\frac{T(T+1)}{2}$. Correspondingly, the number of parameters of the working covariance matrix is reduced to $\frac{T}{8}(T+1)(T^2+T+2)$ from $\frac{T}{8}(T+1)(T^2+5T+6)$. The algorithm for parameter estimation is, however, not substantially changed because fourth-order moments are calculated from first- and second-order moments. The asymptotic properties of this special case of PML2 estimation can be obtained by applying the continuous mapping theorem to the results of Theorem 7.2.

The same principle for estimation can be used for count data. Here, the log link yielding $\mathbb{E}(y_{it}|\boldsymbol{x}_{it}) = \mu_{it} = \exp(\boldsymbol{x}_{it}'\boldsymbol{\beta}_0)$ is the natural choice. Using the assumption $\sigma_{it}^2 = \mu_{it}$ and the model for the correlation coefficient from above, specification of both first- and second-order moments is complete.

Again, we stress that the correct specification of first- and second-order moments is required for consistent estimation of $\boldsymbol{\beta}$ and $\boldsymbol{\alpha}$. Subsequently, if over- or underdispersion is present, the simple functional relation between the mean and the variance does not hold anymore. $\sigma_{it}^2 = v_{it} = \phi h(\mu_{it})$ is often used in these cases instead. In applications, ϕ is estimated either by simple method of moments or by a third set of estimating equations (compare previous section).

7.3.3 Generalized estimating equations 2 with an arbitrary quadratic exponential family using the second centered moments

In Sects. 7.3.1 and 7.3.2 the working covariance matrix was automatically fixed by the normal distribution. Third-order moments were 0, and fourth-order moments were products and sums of second-order moments. This has two specific disadvantages for applications. First, if the variances are estimated to be close to zero, the product of these may be even closer to zero, and inversion of the working covariance matrix could be numerically unstable. This, in turn, can lead to substantial convergence problems of the algorithm for solving the GEE2.

Second, investigators might wish to choose specific structures for the working covariance structure to improve efficiency when the true underlying distribution is not close to the normal distribution. To overcome these problems, Prentice and Zhao (1991) proposed to consider an arbitrary quadratic exponential family using the second centered moments as the measure of association. Specifically, the mean structure may be chosen as in Sect. 7.3.1, i.e., $\mathbb{E}(y_{it}|\boldsymbol{x}_{it}) = \mu_{it} = g(\boldsymbol{x}_{it}'\boldsymbol{\beta}_0)$. Furthermore, the variance is modeled using Eq. 7.2, i.e., $\mathbb{V}\text{ar}(y_{it}|\boldsymbol{X}_i) = \sigma_{it}^2 = \exp(\boldsymbol{x}_{it}'\boldsymbol{\alpha})$. Finally, the covariances are modeled via $\sigma_{itt'} = \mathbb{C}\text{orr}(y_{it}, y_{it'})/\big(\sigma_{it}\sigma_{it'}\big)$, and Fisher's z association link is chosen:

$$\mathbb{C}\text{orr}(y_{it}, y_{it'}) = \frac{\exp\big(\boldsymbol{k}(\boldsymbol{x}_{it}, \boldsymbol{x}_{it'})'\boldsymbol{\alpha})\big) - 1}{\exp\big(\boldsymbol{k}(\boldsymbol{x}_{it}, \boldsymbol{x}_{it'})'\boldsymbol{\alpha})\big) + 1}\,.$$

This completes specification of the first two moments.

For estimation, the PML2 estimating equations based on the score vector from Theorem 7.2, 6. are used. Therefore, we finally have to choose a specific working covariance matrix for third- and fourth-order moments. Below, several working covariance structures are given.

Example 7.5 (Independence working covariance matrix). A simple choice is to assume independence of observations $\boldsymbol{y}_i$ (Prentice and Zhao, 1991). Then, $\mathbb{C}\text{ov}(\boldsymbol{y}_i, \boldsymbol{s}_i) = \boldsymbol{0}$, and $\mathbb{V}\text{ar}(\boldsymbol{s}_i)$ is diagonal with elements $\mathbb{V}\text{ar}(s_{i,tt'}) = \mathbb{V}\text{ar}\big((y_{it} - \mu_{it})^2(y_{it'} - \mu_{it'})^2\big) = \sigma_{it}^2\sigma_{it'}^2$ for $i \neq j$. By using the value from the normal distribution for $\mathbb{V}\text{ar}(s_{i,tt}) = \mathbb{V}\text{ar}\big((y_{it} - \mu_{it})^4\big) = 2\sigma_{it}^2$, the working covariance matrix is completely specified.

This independence working covariance matrix should be chosen only if there is only a weak dependence between y_{it} and $y_{it'}$. Furthermore, a substantial weakness of the independence working covariance matrix is that it my be close to singularity if some values σ_{it}^2 are close to zero. We therefore prefer the working covariance matrix for applications over the independence covariance matrix. This working covariance structure has the flavor of the working covariance matrix from Sect. 5.3.4.

Example 7.6 (Working covariance matrix for applications). The simplest choice of the working variance matrix is (Ziegler et al., 1998)

$$\mathbb{C}\text{ov}(\boldsymbol{y}_i, \boldsymbol{s}_i) = \boldsymbol{0}\,, \qquad \text{and} \qquad \mathbb{V}\text{ar}(\boldsymbol{s}_i) = \boldsymbol{I}\,.$$

This working covariance matrix has two advantages. First, it guarantees regularity of the lower part of the working covariance matrix. It thus avoids convergence problems of GEE2 algorithms. Second, third- and fourth-order moments need not be estimated, which increases the speed of the GEE2 algorithm.

Example 7.7 (Common working correlation of third- and fourth-order moments). This working covariance matrix is a natural generalization of the working covariance matrix under normality. Let

$$\begin{aligned}
\mathbb{C}\text{ov}(y_{it}, s_{i,rr'}) &= \mathbb{E}\big((y_{it} - \mu_{it})(y_{ir} - \mu_{ir})(y_{ir'} - \mu_{ir'})\big) = \gamma_{trr'}(\sigma_{it}^2\sigma_{ir}^2\sigma_{ir'}^2)^{1/2} \\
\mathbb{C}\text{ov}(s_{i,tt'}, s_{i,rr'}) &= \mathbb{E}\big((y_{it} - \mu_{it})(y_{it'} - \mu_{it'})(y_{ir} - \mu_{ir})(y_{ir'} - \mu_{ir'})\big) \\
&\quad -\sigma_{i,tt'}\sigma_{i,rr'}\,, \\
&= \sigma_{i,tr}\sigma_{i,t'r'} + \sigma_{i,tr'}\sigma_{i,t'r} + \delta_{tt'rr'}\sqrt{\sigma_{it}^2\sigma_{it'}^2\sigma_{ir}^2\sigma_{ir'}^2}\,,
\end{aligned}$$

with additional parameters $\gamma_{trr'}$ and $\delta_{tt'rr'}$, which can be estimated consistently using means.

This working covariance matrix can be used to account for skewness and kurtosis of the true distribution, which may deviate from the normal distribution. Hence, this working covariance structure will have good efficiency properties. However, convergence problems generally arise because of sparseness of the data. A simpler version of this covariance matrix can be obtained by equating specific parameters $\gamma_{trr'}$ and $\delta_{tt'rr'}$. This is plausible, if some elements of $\boldsymbol{y}_i$ follow an exchangeable structure.

For dichotomous dependent variables, Prentice and Zhao (1991) proposed the following generalization of the independence working covariance structure.

Example 7.8 (Independence with structural non zeros).

$$\begin{aligned}\mathbb{C}\text{ov}(y_{it}, s_{i,tt'}) &= (1 - 2\mu_{it})\sigma_{i,tt'} \,, \\ \mathbb{V}\text{ar}(s_{i,tt'}) &= \sigma_{it}^2\sigma_{it'}^2 - \sigma_{i,tt'}^2 + (1 - 2\mu_{it})(1 - 2\mu_{it'})\sigma_{i,tt'} \,, \\ \mathbb{C}\text{ov}(s_{i,tt'}, s_{i,tr}) &= \sigma_{it}^2\sigma_{it'r} - \sigma_{i,tt'}\sigma_{i,tr} \,.\end{aligned}$$

The major concern against the use of this working covariance structure in applications is that it may result in a singular working covariance matrix for two reasons. First, both $\mathbb{V}\text{ar}(s_{i,tt'})$ and $\mathbb{C}\text{ov}(s_{i,tt'}, s_{i,tr})$ include the product of two variances, and this product can be close to 0. Second, one term is subtracted for both $\mathbb{V}\text{ar}(s_{i,tt'})$ and $\mathbb{C}\text{ov}(s_{i,tt'}, s_{i,tr})$. This can again lead to values close to 0. Furthermore, because of numeric instability, estimates of $\mathbb{V}\text{ar}(s_{i,tt'})$ can be negative.

As a consequence, we have generally used the working covariance matrix for applications in our own analyses.

7.3.4 Generalized estimating equations 2 for binary data using the second ordinary moments

In the last three sections, the covariance was used as the measure of association. However, the functional relationship between the second centered moments and the association parameter vector was established through a transformation to the correlation coefficient. Specifically, the correlation coefficient was defined as a function of the association parameter $\boldsymbol{\alpha}$, but the covariance $\sigma_{itt'}$ turned out to be a function of both $\boldsymbol{\beta}$ and $\boldsymbol{\alpha}$. An alternative approach is to use the second ordinary moments and the odds ratio (OR) as measure of association. This idea was proposed by Lipsitz et al. (1991), who introduced ad hoc estimating equations, in which the possible dependency of the association structure on $\boldsymbol{\beta}$ was ignored. The approach of Lipsitz et al. (1991) has been extended by Liang et al. (1992) and Qaqish and Liang (1992) to a full GEE2 approach. The OR as the measure of association has two advantages over the correlation coefficient as the measure of association. The association link function can be established in a more natural way. Furthermore, for $T = 2$, the parameter space of the OR is not restricted, and higher order restrictions are less severe than the restriction of the parameter space of the correlation coefficient. We stress, however, that all moments of order three and above are set to 0 in the quadratic exponential family.

If the (log) OR is used as the measure of association, the estimating equations considered of Theorem 7.2, 3. are used to formulate the GEE2 with the simplification that, in analogy to Sect. 7.3.4, the variances are not modeled. The mean structure is chosen using a link function from the GLM, i.e., $\mathbb{E}(y_{it}|\boldsymbol{x}_{it}) = \mu_{it} = F(\boldsymbol{x}_{it}'\boldsymbol{\beta}_0)$, and the variance is modeled using the binomial distribution, i.e., $\mathbb{V}\mathrm{ar}(y_{it}|\boldsymbol{X}_i) = \sigma_{it}^2 = \mu_{it}(1-\mu_{it})$.

The second ordinary moments $\nu_{itt'} = \mathbb{E}(y_{it}y_{it'}|\boldsymbol{X}_i)$ are connected with the OR $\tau_{itt'}$ using Eq. 2.7. Therefore, the association link function can be established by defining $\boldsymbol{\alpha}$ as a linear function of the log odds ratio $\ln(\tau_{itt'})$:

$$\ln(\tau_{itt'}) = \boldsymbol{k}(\boldsymbol{x}_{it}, \boldsymbol{x}_{it'})'\boldsymbol{\alpha}\,.$$

As before, $\boldsymbol{k}$ is a function that correctly specifies the influence of the independent variables on the log OR. This completes specification of the first two moments.

Because $\nu_{itt'}$ is a function of the OR and the marginal means μ_{it} and $\mu_{it'}$, $\nu_{itt'}$ depends on both $\boldsymbol{\beta}$ and $\boldsymbol{\alpha}$. Subsequently, $\partial\boldsymbol{\nu}_i/\partial\boldsymbol{\beta} \neq \boldsymbol{0}$ so that the GEE2 from Theorem 7.2, 3. have to be solved simultaneously.

To complete specification of the GEE2, the working covariance matrix needs to be chosen.

Example 7.9 (Working covariance matrix using properties for dichotomous random variables). Because y_{it} and $w_{itt'}$ are both dichotomous random variables with mean $\pi_{it} = \mu_{it}$ and $\nu_{itt'}$, third-order moments are given by

$$\mathbb{C}\mathrm{ov}(y_{is}, w_{itt'}) = \mathbb{P}(y_{is} = y_{it} = y_{it'} = 1) - \mu_{is}\nu_{itt'}\,,$$

and fourth-order moments are given by

$$\begin{aligned}\mathbb{V}\mathrm{ar}(w_{itt'}) &= \nu_{itt'}(1-\nu_{itt'})\\ \mathbb{C}\mathrm{ov}(w_{itt'}, w_{iss'}) &= \mathbb{P}(y_{is} = y_{is'} = y_{it} = y_{it'} = 1) - \nu_{itt'}\nu_{iss'}\,.\end{aligned}$$

For estimating $\mu_{i,stt'} = \mathbb{P}(y_{is} = y_{it} = y_{it'} = 1)$ and $\mu_{i,ss'tt'} = \mathbb{P}(y_{is} = y_{is'} = y_{it} = y_{it'} = 1)$, either an iterative proportional fitting algorithm (IPF; Heagerty and Zeger, 1996) or a Newton-Raphson type algorithm (Agresti, 1990, p. 188) can be used.

It is questionable whether the effort for estimating $\mu_{i,stt'}$ and $\mu_{i,ss'tt'}$ through CPU time-intensive algorithms is worthwhile because third- and fourth-order moments are nuisance. Simplified working covariance matrices might therefore be preferable in applications.

Example 7.10 (Diagonal working covariance matrix). To increase stability of the working covariance matrix, third-order moments, i.e., the lower left block of the working covariance matrix, are often set to $\mathbf{0}$. Furthermore, the lower right block is chosen to be diagonal with elements $\nu_{itt'}(1 - \nu_{itt'})$.

———

A disadvantage of choosing $\nu_{itt'}(1 - \nu_{itt'})$ as diagonal elements is that these can be close to 0 so that the working covariance matrix is close to singularity. To avoid convergence problems of the algorithm, the working covariance matrix for applications is a reasonable choice for GEE2 with the second centered moments as the measure of association. This choice also has the additional advantage that no further estimations are required.

Example 7.11 (Matrix for applications). Third-order moments are set to 0 for the working covariance matrix for applications. The lower right block of the working covariance matrix representing fourth-order moments is the identity matrix.

———

Further choices of the working covariance matrix are possible. Their value is, however, questionable.

Chapter 8
Generalized method of moment estimation

The generalized method of moments (GMM) was introduced by Hansen in 1982. It is of great importance in econometrics because it provides a unified framework for the analysis of many well-known estimators, such as least squares, instrumental variables (IV), and maximum likelihood (ML). Several excellent book chapters and textbooks are available (Hall, 1993; Ogaki, 1993). Here, we restrict our attention to the elements of GMM theory essential for deriving the generalized estimating equations (GEE).

Therefore, some properties of GMM are derived first, and, second, some special GMM estimators are derived that are equivalent to different GEE2 estimators. In the next section, the GMM estimator as required here is defined.

8.1 Definition

In the previous chapters, we have considered likelihoods – either true or assumed. GMM is not based on likelihoods at all but on moment conditions. To give a simple example, we consider the simple linear model $\boldsymbol{y} = \boldsymbol{X}\boldsymbol{\beta} + \boldsymbol{\varepsilon}$, where $\boldsymbol{y}$ is the $n \times 1$ vector of dependent variables, $\boldsymbol{X}$ is the $n \times p$ matrix of fixed or stochastic independent variables, $\boldsymbol{\beta}_0$ is the true $p \times 1$ parameter vector of interest, and $\boldsymbol{\varepsilon}$ is the $n \times 1$ vector of errors.

The estimator $\hat{\boldsymbol{\beta}}$ is obtained by solving the normal equations $\boldsymbol{X}'\boldsymbol{X}\hat{\boldsymbol{\beta}} = \boldsymbol{X}'\boldsymbol{y}$, which can also be written as $\boldsymbol{X}'(\boldsymbol{X}\hat{\boldsymbol{\beta}} - \boldsymbol{y}) = \boldsymbol{X}'\hat{\boldsymbol{\varepsilon}} = \boldsymbol{0}$. This means that the estimated error $\hat{\boldsymbol{\varepsilon}}$ is orthogonal to the vector space spanned by the design matrix $\boldsymbol{X}$. This property characterizes the linear model, and it can be made an essential moment condition: $\mathbb{E}^{\boldsymbol{X}}\mathbb{E}^{\boldsymbol{y}}\big(\boldsymbol{X}'\boldsymbol{\varepsilon}\big) = \boldsymbol{0}$.

For the general linear model with weight matrix $\boldsymbol{\Sigma}$, this moment condition can be generalized to $\mathbb{E}^{\boldsymbol{X}}\mathbb{E}^{\boldsymbol{y}}\big(\boldsymbol{X}'\boldsymbol{\Sigma}^{-1}\boldsymbol{\varepsilon}\big) = \boldsymbol{0}$, and for the generalized linear model (GLM), it is $\mathbb{E}^{\boldsymbol{X}}\mathbb{E}^{\boldsymbol{y}}\big(\boldsymbol{D}'\boldsymbol{\Sigma}^{-1}\boldsymbol{\varepsilon}\big) = \boldsymbol{0}$, where $\boldsymbol{D} = \partial\boldsymbol{\mu}/\partial\boldsymbol{\beta}'$.

For the general case, we consider the $T \times 1$ vector $\boldsymbol{y}_i$ of dependent variables, the $T \times p$ matrix $\boldsymbol{X}_i$ of independent variables, and an $r \times 1$ parameter of interest $\boldsymbol{\xi}_0$. Note that we consider $\boldsymbol{\xi}$, which generally is the stacked vector of $(\boldsymbol{\beta}', \boldsymbol{\alpha}')'$, where $\boldsymbol{\beta}$ is the $p \times 1$ vector for the mean structure and $\boldsymbol{\alpha}$ the $q \times 1$ vector of the association structure.

We assume that for $i = 1, \ldots, n$, moment conditions can be established and written as

$$\mathbb{E}^{\boldsymbol{X}} \mathbb{E}^{\boldsymbol{y}} \big(\psi(\boldsymbol{y}_i, \boldsymbol{X}_i, \boldsymbol{\xi}_0)\big) = \mathbf{0}$$

for some continuous $r \times 1$ function ψ. Note that we restrict our attention to just identified models so that the parameter vector $\boldsymbol{\xi}$ has r components, and the number of moment conditions is r, too. As before, the pairs $(\boldsymbol{y}_i, \boldsymbol{X}_i)$ are assumed to be independently identically distributed. With this notation, we can define the GMM estimator.

Definition 8.1. A GMM estimator is any value $\hat{\boldsymbol{\xi}}$ minimizing

$$\left(\frac{1}{n}\sum_{i=1}^{n} \psi(\boldsymbol{y}_i, \boldsymbol{X}_i, \boldsymbol{\xi})\right)' \left(\frac{1}{n}\sum_{i=1}^{n} \psi(\boldsymbol{y}_i, \boldsymbol{X}_i, \boldsymbol{\xi})\right) = \psi(\boldsymbol{\xi})'\psi(\boldsymbol{\xi}) \,. \tag{8.1}$$

The reader should note that the GMM estimator is defined in a simple way in Definition 8.1. Specifically, no positive definite weight matrix $\boldsymbol{W}$ is involved that would give $\psi(\boldsymbol{\xi})'\boldsymbol{W}\psi(\boldsymbol{\xi})$ because we consider only just identified models. For the general case, the reader may refer to the literature (see, e.g., Hall, 1993; Ogaki, 1993). Furthermore, we already note that the function ψ is identical to the score vector from the previous chapters.

8.2 Asymptotic properties

In this section, the asymptotic properties of GMM estimators are formulated. The required regularity conditions and detailed proofs can be found, e.g., in Gourieroux and Monfort (1995a, p. 313) and Hansen (1982). To simplify notation, $\partial f(\hat{\boldsymbol{\xi}})/\partial\boldsymbol{\xi}$ means that the first derivative of $f(\boldsymbol{\xi})$ is taken with respect to $\boldsymbol{\xi}$, and the functional is then evaluated at $\hat{\boldsymbol{\xi}}$.

Theorem 8.2.

1. *There exists a GMM estimator* $\hat{\boldsymbol{\xi}}$ *for* $\boldsymbol{\xi}_0$.
2. *The GMM estimator* $\hat{\boldsymbol{\xi}}$ *converges almost surely to the true parameter vector* $\boldsymbol{\xi}_0$.
3. *The GMM estimator* $\hat{\boldsymbol{\xi}}$ *can be obtained by solving the first-order conditions*

$$\boldsymbol{u}(\hat{\boldsymbol{\xi}}) = \frac{1}{n}\sum_{i=1}^{n} \boldsymbol{u}_i(\hat{\boldsymbol{\xi}}) = \frac{1}{n}\sum_{i=1}^{n} \psi_i(\hat{\boldsymbol{\xi}}) = \frac{1}{n}\sum_{i=1}^{n} \psi(\boldsymbol{y}_i, \boldsymbol{X}_i, \hat{\boldsymbol{\xi}}) = \boldsymbol{0}\,.$$

4. *The GMM estimator $\hat{\boldsymbol{\xi}}$ for $\boldsymbol{\xi}_0$ is asymptotically normal. More specifically,*

$$\sqrt{n}(\hat{\boldsymbol{\xi}} - \boldsymbol{\xi}_0) \overset{a}{\sim} N\Big(\boldsymbol{0}, \big(\boldsymbol{A}(\boldsymbol{\xi}_0)\big)^{-1}\boldsymbol{B}(\boldsymbol{\xi}_0)\big(\boldsymbol{A}(\boldsymbol{\xi}_0)'\big)^{-1}\Big)\,,$$

   where

$$\boldsymbol{A}(\boldsymbol{\xi}_0) = \mathbb{E}^{\boldsymbol{X}}\Big(\mathbb{E}^{\boldsymbol{y}}\frac{\partial \boldsymbol{u}_i(\boldsymbol{\xi})}{\partial \boldsymbol{\xi}'}\Big) \text{ and } \boldsymbol{B}(\boldsymbol{\xi}_0) = \mathbb{E}^{\boldsymbol{X}}\Big(\mathbb{E}^{\boldsymbol{y}}\big(\boldsymbol{u}_i(\boldsymbol{\xi})\boldsymbol{u}_i(\boldsymbol{\xi})'\big)\Big)$$

   are the matrices for formulating the sandwich estimator of variance.
5. *Strongly consistent estimators $\boldsymbol{A}(\boldsymbol{\xi}_0)$ and $\boldsymbol{B}(\boldsymbol{\xi}_0)$ are obtained by replacing $\boldsymbol{\xi}_0$ with its estimator.*
6. *There exists a best GMM estimator. It is obtained when $\boldsymbol{A}(\boldsymbol{\xi}_0) = \mathbb{V}ar(\psi_i)$.*

Remark 8.3. The formulation of the covariance matrix of the GMM estimator in Theorem 8.2 seems to be different from the covariance matrix in the PML framework. Specifically, the GMM covariance estimator is of a sandwich form, the matrix $\boldsymbol{A}$ may, however, be non-symmetric. Despite the difference in form, the covariance matrices for GEE estimators derived from PML or GMM are identical, and the matrix $\boldsymbol{A}$ is symmetric in all examples considered here. Therefore, the covariance matrix of the estimators will not be derived in this chapter.

Proof.
1.: Existence: See Hansen (1982, Theorem 2.1).
2.: Consistency: See Hansen (1982, Theorem 2.1).
3.: First-order condition: This is a direct consequence by derivating Eq. 8.1 with respect to $\boldsymbol{\xi}$.
4.: Asymptotic normality: The proof is straightforward. We use the first-order conditions and take a Taylor series expansion around $\boldsymbol{\xi}_0$. We solve for $\sqrt{n}(\hat{\boldsymbol{\xi}} - \boldsymbol{\xi}_0)$, and replace estimated averages with their probability limits. Finally, Cramér's theorem is applied to $\sqrt{n}(\hat{\boldsymbol{\xi}} - \boldsymbol{\xi}_0)$, which completes the proof.

In detail, we have

$$\frac{1}{\sqrt{n}}\sum_{i=1}^{n} \hat{\boldsymbol{u}}_i(\hat{\boldsymbol{\xi}}) = \boldsymbol{0}\,,$$

implying

$$\frac{1}{\sqrt{n}}\sum_{i=1}^{n} \boldsymbol{u}_i(\boldsymbol{\xi}_0) + \frac{1}{n}\sum_{i=1}^{n} \frac{\partial \boldsymbol{u}_i(\boldsymbol{\xi}_0)}{\partial \boldsymbol{\xi}'}\sqrt{n}(\hat{\boldsymbol{\xi}} - \boldsymbol{\xi}_0) \overset{a.s.}{=} \boldsymbol{0}\,.$$

Thus,

$$\sqrt{n}(\hat{\boldsymbol{\xi}} - \boldsymbol{\xi}_0) \stackrel{a.s.}{=} -\mathbb{E}^{\boldsymbol{X}}\mathbb{E}^{\boldsymbol{y}}\Big(\frac{\partial \boldsymbol{u}_i(\boldsymbol{\xi}_0)}{\partial \boldsymbol{\xi}}'\Big)\frac{1}{\sqrt{n}}\sum_{i=1}^{n} \boldsymbol{u}_i(\boldsymbol{\xi}_0).$$

Finally, Cramér's theorem allows us to deduce the desired result.
5.: Estimation: See Hansen (1982, Lemmata 3.2 and 3.3).
6.: Efficiency: This is a direct consequence of Hansen (1982, Theorem 3.2), although Hansen considers the more general case of overidentified GMM with some weight matrix $\boldsymbol{W}$. □

8.3 Examples

To illustrate the strength of the GMM, we consider some examples. A series of applications can be found in the econometric literature, and the broad variety of applications has been nicely summarized in different papers and book chapters (see, e.g., Hall, 2005; Hansen and West, 2002; Jagannathan et al., 2002). The task in GMM is to find the moments that define the function $\psi(\boldsymbol{y}_i, \boldsymbol{X}_i, \boldsymbol{\xi})$. Therefore, emphasis is on the derivation of the minimand and the first-order conditions.

8.3.1 Linear regression

First of all, we consider the classical univariate linear model. The model equation is given by $y_i = \boldsymbol{x}_i'\boldsymbol{\beta} + \varepsilon_i$ for $i = 1, \ldots, n$, where $\boldsymbol{x}_i$ is the p dimensional vector of independent variables, y_i is the dependent variable, ε_i is the residual, and $\boldsymbol{\beta}$ is the parameter of interest.

The well-known ordinary least squares estimation can be embedded into the GMM framework by choosing

$$\psi(\boldsymbol{y}_i, \boldsymbol{X}_i, \boldsymbol{\xi}) = \boldsymbol{x}_i(y_i - \boldsymbol{x}_i'\boldsymbol{\beta}),$$

where $\boldsymbol{\xi} = \boldsymbol{\beta}$. The moment conditions fulfill the orthogonality conditions $\mathbb{E}(\boldsymbol{x}_i\varepsilon_i) = \boldsymbol{0}$, and the GMM estimator minimizes

$$\Big(\frac{1}{n}\sum_{i=1}^{n} \boldsymbol{x}_i(y_i - \boldsymbol{x}_i'\boldsymbol{\beta})\Big)'\Big(\frac{1}{n}\sum_{i=1}^{n} \boldsymbol{x}_i(y_i - \boldsymbol{x}_i'\boldsymbol{\beta})\Big),$$

which is formally different from the ordinary least squares estimator that minimizes $\frac{1}{n}\sum_{i=1}^{n}(y_i - \boldsymbol{x}_i'\boldsymbol{\beta})^2$. Of course, the resulting estimator coincides with the GMM estimator in this case.

8.3.2 Independence estimating equations with covariance matrix equal to identity matrix

We reconsider the example from Sect. 5.3.4. Specifically, we formulate the independence estimating equations (IEE) with identity covariance matrix using the GMM method.

We consider the T-dimensional random vector $\boldsymbol{y}_i = (y_{i1}, \dots, y_{iT})'$, and its associated matrix of explanatory variables $\boldsymbol{X}_i = (\boldsymbol{x}_{i1}, \dots, \boldsymbol{x}_{iT})'$, for $i = 1, \dots, n$. The pairs $(\boldsymbol{y}_i, \boldsymbol{X}_i)$ are assumed to be independent and identically distributed with mean structure $\mathbb{E}(\boldsymbol{y}_i|\boldsymbol{X}_i||\boldsymbol{\beta}_0) = g(\boldsymbol{X}_i\boldsymbol{\beta}_0)$ for a response function g, which is defined element-wise as in multivariate GLM. We furthermore consider a covariance matrix $\boldsymbol{I} = \boldsymbol{I}_{T\times T}$, i.e., $\mathbb{V}\text{ar}(\boldsymbol{y}_i|\boldsymbol{X}_i) = \boldsymbol{I}$.

A natural choice therefore is

$$\psi(\boldsymbol{y}_i, \boldsymbol{X}_i, \boldsymbol{\xi}) = \boldsymbol{D}_i'\big(\boldsymbol{y}_i - g(\boldsymbol{X}_i\boldsymbol{\beta})\big) = \boldsymbol{D}_i'\boldsymbol{\varepsilon}_i\,,$$

where $\boldsymbol{\xi} = \boldsymbol{\beta}$, $\boldsymbol{D}_i = \partial\boldsymbol{\mu}_i/\partial\boldsymbol{\beta}'$ is the matrix of first derivatives, and $\boldsymbol{\varepsilon}_i = \boldsymbol{y}_i - \boldsymbol{\mu}_i = \boldsymbol{y}_i - g(\boldsymbol{X}_i\boldsymbol{\beta})$ is the first-order residual.

The GMM estimator minimizes

$$\Big(\frac{1}{n}\sum_{i=1}^{n}\boldsymbol{D}_i'\boldsymbol{\varepsilon}_i\Big)'\Big(\frac{1}{n}\sum_{i=1}^{n}\boldsymbol{D}_i'\boldsymbol{\varepsilon}_i\Big)\,,$$

yielding the IEE with identity covariance matrix

$$\boldsymbol{u}(\hat{\boldsymbol{\beta}}) = \frac{1}{n}\sum_{i=1}^{n}\hat{\boldsymbol{D}}_i'\hat{\boldsymbol{\varepsilon}}_i = \boldsymbol{0}\,,$$

where $\boldsymbol{D}_i = \partial\boldsymbol{\mu}_i/\partial\boldsymbol{\beta}'$ is the matrix of first derivatives, and $\boldsymbol{\varepsilon}_i = \boldsymbol{y}_i - \boldsymbol{\mu}_i = \boldsymbol{y}_i - g(\boldsymbol{X}_i\boldsymbol{\beta})$ is the first-order residual. These estimating equations are termed IEE with identity covariance matrix.

According to Theorem 8.2, the resulting estimator $\hat{\boldsymbol{\beta}}$ is asymptotically normally distributed. The Fisher information matrix $\boldsymbol{A}$ and the OPG $\boldsymbol{B}$ can be estimated (strongly) consistently as described in Sect. 5.3.4.

8.3.3 Generalized estimating equations 1 with fixed working covariance matrix

Now, we reconsider the example from Sect. 5.3.5 for generalized estimating equations 1 (GEE1) with a fixed covariance matrix. The model is identical to the one in the previous section, but we use an arbitrary fixed covariance matrix $\boldsymbol{\Sigma}_i$ instead of the identity matrix.

For $\boldsymbol{\xi} = \boldsymbol{\beta}$, the function ψ is defined by

$$\psi(\boldsymbol{y}_i, \boldsymbol{X}_i, \boldsymbol{\xi}) = \boldsymbol{D}_i' \boldsymbol{\Sigma}_i^{-1} (\boldsymbol{y}_i - g(\boldsymbol{X}_i \boldsymbol{\beta})) = \boldsymbol{D}_i' \boldsymbol{\Sigma}_i^{-1} \boldsymbol{\varepsilon}_i \,,$$

and the GMM estimator minimizes

$$\Big(\frac{1}{n}\sum_{i=1}^{n} \boldsymbol{D}_i' \boldsymbol{\Sigma}_i^{-1} \boldsymbol{\varepsilon}_i\Big)' \Big(\frac{1}{n}\sum_{i=1}^{n} \boldsymbol{D}_i' \boldsymbol{\Sigma}_i^{-1} \boldsymbol{\varepsilon}_i\Big),$$

yielding the GEE with fixed covariance matrix

$$\boldsymbol{u}(\hat{\boldsymbol{\beta}}) = \frac{1}{n}\sum_{i=1}^{n} \hat{\boldsymbol{D}}_i' \boldsymbol{\Sigma}_i^{-1} \hat{\boldsymbol{\varepsilon}}_i = \boldsymbol{0} \,.$$

8.3.4 Generalized estimating equations 1 for dichotomous dependent variables with fixed working correlation matrix

In this example, we consider GEE for dependent dichotomous variables and a fixed working correlation structure. The model is identical to the one in Sect. 8.3.2, but we specifically assume that a cumulative distribution function F is used as response function g, and the variance is from the Bernoulli distribution $\mathbb{V}\text{ar}(y_{it}|\boldsymbol{x}_{it}) = v_{it} = h(\mu_{it}) = \mu_{it}(1 - \mu_{it})$. The conditional variance of y_{it} given $\boldsymbol{x}_{it}$ therefore is independent of an additional nuisance parameter Ψ (compare Sect. 6.3.2). We furthermore partition the covariance matrix $\boldsymbol{\Sigma}_i$ in the diagonal matrix of variances $\boldsymbol{V}_i = \text{diag}(v_{it})$ and a fixed working correlation matrix $\boldsymbol{R}_i$.

For $\boldsymbol{\xi} = \boldsymbol{\beta}$, the function ψ is therefore defined by

$$\psi(\boldsymbol{y}_i, \boldsymbol{X}_i, \boldsymbol{\xi}) = \boldsymbol{D}_i' \boldsymbol{V}_i^{-1/2} \boldsymbol{R}_i^{-1} \boldsymbol{V}_i^{-1/2} (\boldsymbol{y}_i - g(\boldsymbol{X}_i \boldsymbol{\beta})) = \boldsymbol{D}_i' \boldsymbol{V}_i^{-1/2} \boldsymbol{R}_i^{-1} \boldsymbol{V}_i^{-1/2} \boldsymbol{\varepsilon}_i \,,$$

and the GMM estimator minimizes

$$\Big(\frac{1}{n}\sum_{i=1}^{n} \boldsymbol{D}_i' \boldsymbol{V}_i^{-1/2} \boldsymbol{R}_i^{-1} \boldsymbol{V}_i^{-1/2} \boldsymbol{\varepsilon}_i\Big)' \Big(\frac{1}{n}\sum_{i=1}^{n} \boldsymbol{D}_i' \boldsymbol{V}_i^{-1/2} \boldsymbol{R}_i^{-1} \boldsymbol{V}_i^{-1/2} \boldsymbol{\varepsilon}_i\Big),$$

yielding the GEE1 with fixed working correlation matrix

$$\boldsymbol{u}(\hat{\boldsymbol{\beta}}) = \frac{1}{n}\sum_{i=1}^{n} \hat{\boldsymbol{D}}_i' \hat{\boldsymbol{V}}_i^{-1/2} \boldsymbol{R}_i^{-1} \hat{\boldsymbol{V}}_i^{-1/2} \hat{\boldsymbol{\varepsilon}}_i = \boldsymbol{0} \,. \tag{8.2}$$

The important aspect of these GEE is that the variance depends only on the parameter of interest $\boldsymbol{\beta}$, and it is independent of additional nuisance parameters $\boldsymbol{\alpha}$. The estimating equations therefore involve only $\boldsymbol{\beta}$ and the pre-specified, i.e., fixed working correlation matrix, $\boldsymbol{R}_i$.

The estimating equations (8.2) can be solved using GMM because no nuisance parameter needs to be estimated. However, the GEE1 with an estimated working correlation matrix (see Sect. 6.3.3) cannot be embedded into the framework of GMM. Specifically, Breitung and Lechner (1995) embed GEE into GMM, but they only allow the weight matrix $\boldsymbol{\Sigma}_i$ to depend on the mean structure parameter $\boldsymbol{\beta}$. The weight matrix may not depend on an additional nuisance parameter $\boldsymbol{\alpha}$.

8.3.5 Generalized estimating equations 2 for binary data using the second ordinary moments

PML2 estimation can be embedded into the framework of GMM estimation, see, e.g., Ziegler (1995). To this end, we explicitly consider the case $\boldsymbol{\xi} = (\boldsymbol{\beta}'\boldsymbol{\alpha}')'$, and we use the notation from the previous chapter. To formulate the GEE2 using the second ordinary moments, we define the function ψ as

$$\psi(\boldsymbol{y}_i, \boldsymbol{X}_i, \boldsymbol{\xi}) = \boldsymbol{M}_i'\boldsymbol{V}_i^{-1}\boldsymbol{m}_i\,,$$

where $\boldsymbol{M}_i$, $\boldsymbol{V}_i$, and $\boldsymbol{m}_i$ are defined as in Theorem 7.2, i.e.,

$$\tilde{\boldsymbol{M}}_i = \begin{pmatrix} \dfrac{\partial \boldsymbol{\mu}_i}{\partial \boldsymbol{\beta}'} & \boldsymbol{0} \\ \dfrac{\partial \boldsymbol{\nu}_i}{\partial \boldsymbol{\beta}'} & \dfrac{\partial \boldsymbol{\nu}_i}{\partial \boldsymbol{\alpha}'} \end{pmatrix}, \tilde{\boldsymbol{V}}_i = \begin{pmatrix} \boldsymbol{\Sigma}_i & \mathbb{C}\mathrm{ov}(\boldsymbol{y}_i, \boldsymbol{w}_i) \\ \mathbb{C}\mathrm{ov}(\boldsymbol{w}_i, \boldsymbol{y}_i) & \mathbb{V}\mathrm{ar}(\boldsymbol{w}_i) \end{pmatrix}, \tilde{\boldsymbol{m}}_i = \begin{pmatrix} \boldsymbol{y}_i - \boldsymbol{\mu}_i \\ \boldsymbol{w}_i - \boldsymbol{\nu}_i \end{pmatrix},$$

with $\boldsymbol{s}_i = (s_{i11}, s_{i12}, \ldots, s_{iTT})'$, $\boldsymbol{\sigma}_i = (\sigma_{i11}, \sigma_{i12}, \ldots, \sigma_{iTT})'$, $s_{itt'} = (y_{it} - \mu_{it})(y_{it'} - \mu_{it'})$, and $\sigma_{itt'} = \nu_{itt'} - \mu_{it}\mu_{it'}$. $\boldsymbol{V}_i$ is the working covariance matrix consisting of correctly specified second-order moments $\boldsymbol{\Sigma}_i$ and possibly misspecified thir- and fourth-order moments.

The GMM estimator minimizes

$$\Big(\frac{1}{n}\sum_{i=1}^{n} \boldsymbol{M}_i'\boldsymbol{V}_i^{-1}\boldsymbol{m}_i\Big)'\Big(\frac{1}{n}\sum_{i=1}^{n} \boldsymbol{M}_i'\boldsymbol{V}_i^{-1}\boldsymbol{m}_i\Big),$$

yielding the GEE2 using the second ordinary moments

$$\boldsymbol{u}(\hat{\boldsymbol{\xi}}) = \frac{1}{n}\sum_{i=1}^{n} \hat{\boldsymbol{M}}_i'\hat{\boldsymbol{V}}_i^{-1}\hat{\boldsymbol{m}}_i = \boldsymbol{0}\,.$$

8.3.6 Generalized estimating equations 2 using the second centered moments

Similarly, we can derive the GEE2 using the second centered moments. Here, the function ψ is given by

$$\psi(\boldsymbol{y}_i, \boldsymbol{X}_i, \boldsymbol{\xi}) = \tilde{\boldsymbol{M}}_i' \tilde{\boldsymbol{V}}_i^{-1} \tilde{\boldsymbol{m}}_i \,,$$

where

$$\tilde{\boldsymbol{M}}_i = \begin{pmatrix} \dfrac{\partial \boldsymbol{\mu}_i}{\partial \boldsymbol{\beta}'} & \boldsymbol{0} \\ \dfrac{\partial \boldsymbol{\nu}_i}{\partial \boldsymbol{\beta}'} & \dfrac{\partial \boldsymbol{\nu}_i}{\partial \boldsymbol{\alpha}'} \end{pmatrix}, \tilde{\boldsymbol{V}}_i = \begin{pmatrix} \boldsymbol{\Sigma}_i & \mathbb{C}\mathrm{ov}(\boldsymbol{y}_i, \boldsymbol{w}_i) \\ \mathbb{C}\mathrm{ov}(\boldsymbol{w}_i, \boldsymbol{y}_i) & \mathbb{V}\mathrm{ar}(\boldsymbol{w}_i) \end{pmatrix}, \tilde{\boldsymbol{m}}_i = \begin{pmatrix} \boldsymbol{y}_i - \boldsymbol{\mu}_i \\ \boldsymbol{w}_i - \boldsymbol{\nu}_i \end{pmatrix},$$

with $\nu_{itt'} = \mathbb{E}(y_{it} y_{it'} | \boldsymbol{X}_i)$, $\boldsymbol{\nu}_i = (\nu_{i11}, \nu_{i12}, \ldots, \nu_{iTT})'$, and $\boldsymbol{w}_i = (y_{i1}^2, y_{i1}y_{i2}, \ldots, y_{i1}y_{iT}, y_{i2}^2, y_{i2}y_{i3}, \ldots, y_{iT}^2)'$. $\tilde{\boldsymbol{V}}_i$ is the working covariance matrix consisting of the correctly specified second-order moments $\boldsymbol{\Sigma}_i$ and possibly misspecified third- and fourth-order moments.

The GMM estimator is obtained by minimizing

$$\left(\frac{1}{n} \sum_{i=1}^{n} \tilde{\boldsymbol{M}}_i' \tilde{\boldsymbol{V}}_i^{-1} \tilde{\boldsymbol{m}}_i \right)' \left(\frac{1}{n} \sum_{i=1}^{n} \tilde{\boldsymbol{M}}_i' \tilde{\boldsymbol{V}}_i^{-1} \tilde{\boldsymbol{m}}_i \right),$$

and the first-order conditions

$$\boldsymbol{u}(\hat{\boldsymbol{\xi}}) = \frac{1}{n} \sum_{i=1}^{n} \hat{\tilde{\boldsymbol{M}}}_i' \hat{\tilde{\boldsymbol{V}}}_i^{-1} \hat{\tilde{\boldsymbol{m}}}_i = \boldsymbol{0}$$

are the GEE2 using the second centered moments as the measure of association.

8.3.7 Generalized estimating equations 2 using the second standardized moments

GMM estimation has one advantage compared with PML2 estimation. It is possible to formulate GEE2 using the second standardized moments as the measure of association as shown by Prentice (1988). To this end, we introduce additional notation. Let

$$z_{itt'} = \frac{(y_{it} - \mu_{it})}{\sigma_{it}} \frac{(y_{it'} - \mu_{it'})}{\sigma_{it'}}$$

be the sample correlation coefficient (Prentice, 1988) and $\varrho_{itt'}$ its expected value. Collect all values to vectors, i.e., $\boldsymbol{z}_i = (z_{i11}, \ldots, z_{iTT})'$ and $\boldsymbol{\varrho}_i = (\varrho_{i11}, \ldots, \varrho_{iTT})'$.

Now, we define $\tilde{\tilde{\boldsymbol{M}}}_i, \tilde{\tilde{\boldsymbol{V}}}_i$, and $\tilde{\tilde{\boldsymbol{m}}}_i$ as follows:

$$\tilde{\tilde{\boldsymbol{M}}}_i = \begin{pmatrix} \dfrac{\partial \boldsymbol{\mu}_i}{\partial \boldsymbol{\beta}'} & \mathbf{0} \\ \mathbf{0} & \dfrac{\partial \boldsymbol{\varrho}_i}{\partial \boldsymbol{\alpha}'} \end{pmatrix}, \tilde{\tilde{\boldsymbol{V}}}_i = \begin{pmatrix} \boldsymbol{\Sigma}_i & \mathbf{0} \\ \mathbf{0} & \mathbb{V}\mathrm{ar}(\boldsymbol{z}_i) \end{pmatrix}, \tilde{\tilde{\boldsymbol{m}}}_i = \begin{pmatrix} \boldsymbol{y}_i - \boldsymbol{\mu}_i \\ \boldsymbol{z}_i - \boldsymbol{\varrho}_i \end{pmatrix}.$$

The function ψ is given by

$$\psi(\boldsymbol{y}_i, \boldsymbol{X}_i, \boldsymbol{\xi}) = \tilde{\tilde{\boldsymbol{M}}}_i' \tilde{\tilde{\boldsymbol{V}}}_i^{-1} \tilde{\tilde{\boldsymbol{m}}}_i ,$$

yielding

$$\left(\frac{1}{n} \sum_{i=1}^{n} \tilde{\tilde{\boldsymbol{M}}}_i' \tilde{\tilde{\boldsymbol{V}}}_i^{-1} \tilde{\tilde{\boldsymbol{m}}}_i \right)' \left(\frac{1}{n} \sum_{i=1}^{n} \tilde{\tilde{\boldsymbol{M}}}_i' \tilde{\tilde{\boldsymbol{V}}}_i^{-1} \tilde{\tilde{\boldsymbol{m}}}_i \right)$$

as minimand. The GEE2 using the second standardized moments are obtained as first-order conditions:

$$\boldsymbol{u}(\hat{\boldsymbol{\xi}}) = \frac{1}{n} \sum_{i=1}^{n} \hat{\tilde{\tilde{\boldsymbol{M}}}}_i' \hat{\tilde{\tilde{\boldsymbol{V}}}}_i^{-1} \hat{\tilde{\tilde{\boldsymbol{m}}}}_i = \mathbf{0} .$$

These GEE2 have a substantial advantage over the GEE2 in the second ordinary moments and the GEE2 in the second centered moments. The matrices $\tilde{\tilde{\boldsymbol{M}}}_i$ and $\tilde{\tilde{\boldsymbol{V}}}_i$ are block diagonal so that the estimating equations can be solved separately. As a result, the numerical complexity reduces substantially because $\boldsymbol{\beta}$ and $\hat{\boldsymbol{\alpha}}$ can be estimated using an alternating algorithm. Given an initial set of values for $\boldsymbol{\beta}$, $\hat{\boldsymbol{\beta}}$ is first estimated using a standard algorithm from the generalized linear model (GLM) neglecting the correlation. Next, $\hat{\boldsymbol{\alpha}}$ is estimated using the estimate $\hat{\boldsymbol{\beta}}$. $\hat{\boldsymbol{\alpha}}$ is then used to update $\hat{\boldsymbol{\beta}}$, etc. Another advantage of separating the estimating equations is that the estimator $\hat{\boldsymbol{\beta}}$ of $\boldsymbol{\beta}_0$ remains consistent even if the second-order moments $\boldsymbol{\varrho}_i$ are misspecified.

The relevant aspect for this important simplification is that the correlation function $\boldsymbol{\varrho}_i$ is independent of the mean structure parameter $\boldsymbol{\beta}$ so that $\partial \boldsymbol{\varrho}_i / \partial \boldsymbol{\beta}' = \mathbf{0}$. This can be established as follows. First, the link function is chosen as in GLM, i.e., $\mathbb{E}(y_{it}|\boldsymbol{x}_{it}) = \mu_{it} = g(\boldsymbol{x}_{it}'\boldsymbol{\beta}_0)$.

The variance is also modeled using GLM, i.e., $\mathbb{V}\mathrm{ar}(y_{it}|\boldsymbol{X}_i) = \sigma_{it}^2 = \varphi h(\mu_{it})$, and it is important to note that a nuisance parameter φ needs to be estimated in several models although it is identical to 1 in standard binomial and Poisson models. This nuisance parameter φ either is estimated using simple moment estimators or may be formulated as a function of independent variables $\boldsymbol{X}_i$; compare Sect. 7.3.1.

The correlation is modeled using Fisher's z association link

$$\mathbb{C}\text{orr}(y_{it}, y_{it'}) = \frac{\exp\big(\boldsymbol{k}(\boldsymbol{x}_{it}, \boldsymbol{x}_{it'})'\boldsymbol{\alpha})\big) - 1}{\exp\big(\boldsymbol{k}(\boldsymbol{x}_{it}, \boldsymbol{x}_{it'})'\boldsymbol{\alpha})\big) + 1},$$

and this function is independent of $\boldsymbol{\beta}$, yielding $\partial\boldsymbol{\varrho}_i/\partial\boldsymbol{\beta}' = \mathbf{0}$.

To complete the specification of the first two moments, the covariances are obtained via $\sigma_{itt'} = \mathbb{C}\text{orr}(y_{it}, y_{it'})/\big(\sigma_{it}\sigma_{it'}\big)$.

Finally, since third-order moments of the working covariance matrix equal 0, a working covariance matrix $\mathbb{V}\text{ar}(\boldsymbol{z}_i)$ needs to be chosen for $\boldsymbol{z}_i$. Here, standard choices are those described in Sect. 7.3 in the context of other GEE2 models.

8.3.8 Alternating logistic regression

As discussed in the previous section, parameter estimates of the mean structure may be biased if the estimating equations of both the mean and the association structure are solved simultaneously. The computational burden is also higher in this case. Therefore, estimating equations for the mean and the association structures that can be solved separately are desirable. One such set of estimating equations has been introduced in the previous section.

It is important to note that the estimating equations can be separated if the matrix of first-order derivatives $\tilde{\boldsymbol{M}}_i$ is block diagonal because the working covariance matrix $\boldsymbol{V}_i$ needs to be block diagonal in this case. Specifically, if $\partial\boldsymbol{\varrho}_i/\partial\boldsymbol{\beta}' = \mathbf{0}$ but $\boldsymbol{V}_{i12} = \mathbb{C}\text{ov}(\boldsymbol{y}_i, \boldsymbol{z}_i) \neq \mathbf{0}$, the score vector for $\boldsymbol{\beta}$ is

$$\frac{1}{n}\sum_{i=1}^{n} \boldsymbol{D}_i'\boldsymbol{\Sigma}_i^{-1}(\boldsymbol{y}_i - \boldsymbol{\mu}_i) + \frac{1}{n}\sum_{i=1}^{n} \boldsymbol{D}_i'\boldsymbol{V}_{i12}(\boldsymbol{z}_i - \boldsymbol{\varrho}_i),$$

so that $\boldsymbol{V}_{i12}$ needs to be $\mathbf{0}$ to guarantee consistency of $\hat{\boldsymbol{\beta}}$. In summary, the estimating equations can be solved separately if the function of the association parameter used in the estimating equation is defined independently of the mean structure parameter.

For dichotomous dependent variables, the standard measure of association is the odds ratio (OR), which has several advantages over the correlation coefficient (Sect. 2.4.3). The idea to form two separate sets of estimating equations using the OR as the measure of association was proposed by Carey et al. (1993), and their idea has first been mentioned in the discussion of the paper of Liang et al. (1992) by Diggle (1992) and Firth (1992). The approach is termed alternating logistic regression (ALR) because one logistic regression is performed for the mean structure, and another one is carried out for the association structure.

ALR is based on the following theorem, which was given by Diggle (1992):

Theorem 8.4. *Let $\tau_{itt'}$ be the OR between y_{it} and $y_{it'}$, $\mu_{it} = I\!P(y_{it}=1)$ and $\nu_{itt'} = I\!E(w_{itt'}) = I\!P(y_{it} = y_{it'} = 1)$. Then,*

$$logit\big(I\!P(y_{it}=1|y_{it'})\big) \approx y_{it'} \ln \tau_{itt'} + \ln\left(\frac{\mu_{it} - \nu_{itt'}}{1 - \mu_{it} - \mu_{it'} + \nu_{itt'}}\right). \qquad (8.3)$$

Proof. Using the definition of the OR

$$\tau_{itt'} = \frac{\mathbb{P}(y_{it} = y_{it'} = 1)\ \mathbb{P}(y_{it} = y_{it'} = 0)}{\mathbb{P}(y_{it} = 0, y_{it'} = 1)\ \mathbb{P}(y_{it} = 1, y_{it'} = 0)}\,,$$

we obtain

$$\begin{aligned}\ln \tau_{itt'} &= \ln \frac{\mathbb{P}(y_{it} = 1|y_{it'} = 1)\ \mathbb{P}(y_{it'} = 1)\ \mathbb{P}(y_{it} = y_{it'} = 0)}{\mathbb{P}(y_{it} = 1|y_{it'} = 1)\ \mathbb{P}(y_{it'} = 1)\ \mathbb{P}(y_{it} = 1, y_{it'} = 0)} \\ &= \text{logit}\ \mathbb{P}(y_{it} = 1|y_{it'} = 1) + \ln\left(\frac{1 - \mu_{it} - \mu_{it'} + \nu_{itt'}}{\mu_{it} - \nu_{itt'}}\right) \qquad (8.4)\end{aligned}$$

for the log OR after simple algebra.

Solving the last expression with respect to $\tau_{itt'}$ yields the desired result for $y_{it'} = 1$. The case $y_{it'} = 0$ is proven analogously. □

Equation 8.3 forms the basis for ALR, and it can be interpreted as follows (Carey et al., 1993). Suppose that $\alpha = \ln \tau_{itt'}$ and consider the second term on the right side of Eq. 8.3 as fixed, i.e., as an offset. Of course, this is a simplification because the offset depends on the current values of $\boldsymbol{\beta}$ and $\boldsymbol{\alpha}$. Then, the pairwise log OR α is the regression coefficient of a logistic regression of y_{it} on $y_{it'}$.

Now, we define the log OR in a more general way as a linear function of $\boldsymbol{\alpha}$. Again, we fix the right side of Eq. 8.3. Then, $\ln \tau_{itt'} = \boldsymbol{k}(\boldsymbol{x}_{it}, \boldsymbol{x}_{it'})'\boldsymbol{\alpha}$ for some fixed function $\boldsymbol{k}$. This representation allows estimation of $\boldsymbol{\alpha}$ by a logistic regression of y_{it} on $y_{it'}\boldsymbol{k}(\boldsymbol{x}_{it}, \boldsymbol{x}_{it'})$.

Therefore, the function $\psi(\boldsymbol{y}_i, \boldsymbol{X}_i, \boldsymbol{\xi})$ consists of two parts. The estimating equations for $\boldsymbol{\beta}$ are given by

$$\mathbf{0} = \sum_{i=1}^{n} \hat{\boldsymbol{D}}_i' \hat{\boldsymbol{\Sigma}}_i^{-1} (\boldsymbol{y}_i - \hat{\boldsymbol{\mu}}_i)\,,$$

and the second set of estimating equations for $\boldsymbol{\alpha}$ is now given by

$$\mathbf{0} = \sum_{i=1}^{n} \left(\frac{\partial \hat{\varsigma}_i}{\partial \boldsymbol{\alpha}'}\right)' \widehat{\mathbb{V}\text{ar}}(\hat{\boldsymbol{\varepsilon}}_i)^{-1} \hat{\boldsymbol{\varepsilon}}_i\,,$$

where $\boldsymbol{\varepsilon}_i = (\varepsilon_{i,12}, \ldots, \varepsilon_{i,(T-1)T})'$ is the $\frac{T(T-1)}{2}$ vector of residuals $\varepsilon_{i,tt'} = y_{it} - \varsigma_{itt'}$.

Thus, the function ψ is given by

$$\psi(\boldsymbol{y}_i, \boldsymbol{X}_i, \boldsymbol{\xi}) = \begin{pmatrix} \boldsymbol{D}_i' \boldsymbol{\Sigma}_i^{-1} (\boldsymbol{y}_i - \boldsymbol{\mu}_i) \\ \dfrac{\partial \boldsymbol{\varsigma}_i}{\partial \boldsymbol{\alpha}'}' \mathbb{V}\text{ar}(\boldsymbol{\varepsilon}_i)^{-1} \boldsymbol{\varepsilon}_i \end{pmatrix} .$$

From this, the minimand can be derived, and the first-order conditions have already been formulated.

An important aspect is the specification of the working covariance matrix $\mathbb{V}\text{ar}(\hat{\boldsymbol{\varepsilon}}_i)$. It can be specified using the variance function of the binomial distribution because $\varepsilon_{itt'}$ is a centered dichotomous random variable. Specifically, the identity working covariance matrix or a diagonal working covariance matrix with elements $\varsigma_{itt'}(1 - \varsigma_{itt'})$ might be chosen.

The asymptotic properties of the resulting estimator $\hat{\boldsymbol{\xi}} = (\hat{\boldsymbol{\beta}}', \hat{\boldsymbol{\alpha}}')'$ can be established using Theorem 8.2. Hence, $\hat{\boldsymbol{\xi}}$ is a consistent estimator of $\boldsymbol{\xi}$ and asymptotically normally distributed.

Again, the separation of the GEE2 in a set of two independent estimating equations results in a substantial speed-up of the algorithm (Carey et al., 1993). The second substantial advantage of ALR is that the parameter estimates of the mean structure $\hat{\boldsymbol{\beta}}$ are consistent for $\boldsymbol{\beta}$ even if the association structure is misspecified. This is in contrast to the GEE2 approach from the previous chapter, which yields consistent parameter estimates $\hat{\boldsymbol{\beta}}$ of $\boldsymbol{\beta}$ only if both the mean and the association structure are correctly specified. Finally, Kuk (2004) has shown that ALR is invariant to permutations of the dependent variables within clusters. He also presents a symmetrized version of the estimating equations so that the standard is also permutation-invariant.

A different approach using similar ideas is that of Firth (1992). He suggested the following algorithm to guarantee consistency of $\hat{\boldsymbol{\beta}}$ even if the second-order moments $\boldsymbol{\nu}_i(\boldsymbol{\alpha})$ are misspecified. In the first step, estimate $\hat{\boldsymbol{\beta}}^{(0)}$ from GEE1 or from IEE. In the second step, use GEE2 with $\boldsymbol{\beta}$ fixed at $\hat{\boldsymbol{\beta}}^{(0)}$ to obtain $\hat{\boldsymbol{\alpha}}$. In the third step, use the upper left block for GEE1 with $\boldsymbol{\alpha}$ fixed at $\hat{\boldsymbol{\alpha}}$ to obtain $\hat{\boldsymbol{\beta}}$. This algorithm follows the idea of QGPML estimation (Chapter 6). However, asymptotic properties of the joint parameter vector estimator $\hat{\boldsymbol{\xi}}$ can be obtained from GMM estimation, where sequential estimation is applied. The suggestion of Firth (1992) has not been used in applications because of its great computational effort. In contrast, ALR is implemented in several software packages and regularly used in GEE2 applications.

8.4 Final remarks

In the last three chapters, we have derived many popular GEE for the mean structure and GEE for both the mean and the association structure. Several other extensions have been proposed for making the GEE more efficient. For example, three estimating equations have been considered in some articles (Lee et al., 2008; Yan and Fine, 2004). Alternative formulations of the estimating equations have also been proposed (see, e.g., Hall and Severini, 1998; also see Sutradhar and Kumar, 2001).

Extensions of GEE have been proposed in many different directions, and they include approaches for dealing with missing data (for an overview, see, e.g., Ziegler et al., 2003), approaches for sample size calculations (reviewed in Dahmen and Ziegler, 2004), regression diagnostics (Preisser and Qaqish, 1996; Venezuela et al., 2007; Vens and Ziegler, 2011), and goodness-of-fit tests (see, e.g., Evans and Hosmer, 2004; Horton et al., 1999). Other methodological questions that have been addressed include hypothesis testing (Rotnitzky and Jewell, 1990) and specific models, such as zero-inflated models for count data (Lee et al., 2011).

References

Agresti, A. *Categorical Data Analysis*. Wiley, New York, 1990.

Anderson, T. *An Introduction to Multivariate Statistical Analysis*, Second Edition. Wiley, New York, 1984.

Antoniou, A., & Lu, W.-S. *Practical Optimization: Algorithms and Engineering Applications*. Springer, New York, 2007.

Arminger, G. Specification and estimation of mean structures. In Arminger, G., Clogg, C., & Sobel, M., editors, *Handbook of Statistical Modeling for the Behavioral Sciences*, pp. 77–183. Plenum, New York, 1995.

Bahadur, R. A representation of the joint distribution of responses to n dichotomous items. In Solomon, H., editor, *Studies in Item Analysis and Prediction*, pp. 158–168. Stanford University Press, Stanford, 1961.

Baradat, P., Maillart, M., Marpeau, A., Slak, M. F., Yani, A., & Pastiszka, P. Utility of terpenes to assess population structure and mating patterns in conifers. In Philippe, B., Thomas, A. W., & Müller-Starck, G., editors, *Population Genetics and Genetic Conservation of Forest Trees*, pp. 5–27. Academic Publishing, Amsterdam, 1996.

Baum, C. F., Schaffer, M. E., & Stillman, S. Instrumental variables and GMM: Estimation and testing. *Stata J*, 3:1–31, 2003.

Binder, D. On the variances of asymptotically normal estimators from complex surveys. *Int Stat Rev*, 51:279–292, 1983.

Bishop, Y., Fienberg, S., & Holland, P. *Discrete Multivariate Analysis: Theory and Practice*. MIT Press, Cambridge, 1975.

Box, G., & Cox, D. R. An analysis of transformations. *J R Stat Soc B*, 26: 211–252, 1964.

Breitung, J., & Lechner, M. GMM-estimation of nonlinear models on panel data. Technical report, SFB Discussion Paper, Humboldt-Universität Berlin, 1995.

Breslow, N. Test of hypotheses in overdispersion regression and other quasi-likelihood models. *J Am Stat Assoc*, 85:565–571, 1990.

Broze, L., & Gourieroux, C. Pseudo-maximum likelihood method, adjusted pseudo-maximum likelihood method and covariance estimators. *J Econometrics*, 85:75–98, 1998. doi:10.1016/S0304-4076(97)00095-X

Burguete, J., Gallant, R., & Souza, G. On unification of the asymptotic theory of nonlinear econometric models. *Economet Rev*, 2:150–190, 1982. doi:10.1080/07311768208800012

Carey, V., Zeger, S. L., & Diggle, P. Modelling multivariate binary data with alternating logistic regression. *Biometrika*, 80:517–526, 1993. doi:10.1093/biomet/80.3.517

Carroll, R., & Ruppert, D. *Transformation and Weighting in Regression*. Chapman and Hall, New York, 1988.

Carroll, R. J., Wang, S., Simpson, D. G., Stromberg, A. J., & Ruppert, D. The sandwich (robust covariance matrix) estimator. Technical report, Department of Statistics, Texas A&M University, 1998.

Chaganty, N. R., & Deng, Y. Ranges of measures of association for familial binary variables. *Commun Stat – Theor M*, 36:587–598, 2007. doi:10.1080/03610920601001808

Cochran, W. G. *Sampling Techniques*, Second Edition. Wiley, New York, 1963.

Cochrane, D., & Orcutt, G. Application of least squares regression to relationships containing autocorrelated terms. *J Am Stat Assoc*, 44:32–61, 1949.

Cook, R. D., & Weisberg, S. *Residuals and Influence in Regression*. Chapman and Hall, New York, 1982.

Cox, D. R., & Reid, N. Parameter orthogonality and approximate conditional inference. *J R Stat Soc B*, 49:1–39, 1987.

Dahmen, G., & Ziegler, A. Generalized estimating equations in controlled clinical trials: Hypotheses testing. *Biom J*, 46:214–232, 2004. doi:10.1002/bimj.200310018

Davis, C. S. Semi-parametric and non-parametric methods for the analysis of repeated measurements with applications to clinical trials. *Stat Med*, 10: 1959–1980, 1991. doi:10.1002/sim.4780101210

Dhaene, G., & Hoorelbeke, D. The information matrix test with bootstrap-based covariance matrix estimation. *Econom Lett*, 82:341–347, 2003. doi:10.1016/j.econlet.2003.09.002

Diggle, P. Discussion of "Multivariate regression analysis for categorical data." *J R Stat Soc B*, 54:28–29, 1992.

Dobson, A. J. *Introduction to Generalized Linear Models*, Second Edition. Chapman and Hall, London, 2001.

Efron, B. Discussion of "Jackknife, bootstrap and other resampling methods in statistics." *Ann Stat*, 14:1301–1304, 1986. doi:10.1214/aos/1176350145

Efron, B., & Hinkley, D. Assessing the accuracy of the maximum likelihood estimation: Observed versus expected information. *Biometrika*, 65:457–482, 1978. doi:10.1093/biomet/65.3.457

Emrich, L. J., & Piedmonte, M. R. On some small sample properties of generalized estimating equation estimates for multivariate dichotomous outcomes. *J Stat Comput Sim*, 41:19–29, 1992. doi:10.1080/00949659208811388

Evans, S. R., & Hosmer, D. W. Goodness of fit tests for logistic GEE models: Simulation results. *Commun Stat – Simul C*, 33:247–258, 2004. doi:10.1081/SAC-120028443

Fahrmeir, L., & Tutz, G. *Multivariate Statistical Modelling Based on Generalized Linear Models*, Second Edition. Springer, New York, 2001.

Fay, M. P., & Graubard, B. I. Small-sample adjustments for Wald-type tests using sandwich estimators. *Biometrics*, 57:1198–1206, 2001. doi:10.1111/j.0006-341X.2001.01198.x

Fay, M. P., Graubard, B. I., Freedman, L. S., & Midthune, D. N. Conditional logistic regression with sandwich estimators: Application to a meta-analysis. *Biometrics*, 54:195–208, 1998.

Firth, D. Discussion of "Multivariate regression analysis for categorical data." *J R Stat Soc B*, 54:24–26, 1992.

Fitzmaurice, G. M., & Laird, N. M. A likelihood-based method for analysing longitudinal binary responses. *Biometrika*, 80:141–151, 1993. doi:10.1093/biomet/80.1.141

Foncel, J., Hristache, M., & Patilea, V. Semiparametric single-index Poisson regression model with unobserved heterogeneity. Working paper 2004–04, Institut National de la Statistique et de Etudes Economiques, Série des Documents de Travail du CREST, 2004.

Gourieroux, C., & Monfort, A. Pseudo-likelihood methods. In Maddala, G., Rao, C., & Vinod, H., editors, *Handbook of Statistics*, Vol. 11, pp. 335–362. Elsevier, Amsterdam, 1993.

Gourieroux, C., & Monfort, A. *Statistics and Econometric Models*, Vol. 1. Cambridge University Press, Cambridge, 1995a.

Gourieroux, C., & Monfort, A. *Statistics and Econometric Models*, Vol. 2. Cambridge University Press, Cambridge, 1995b.

Gourieroux, C., Monfort, A., & Trognon, A. Pseudo maximum likelihood methods: Applications to Poisson models. *Econometrica*, 52:701–720, 1984a.

Gourieroux, C., Monfort, A., & Trognon, A. Pseudo maximum likelihood methods: Theory. *Econometrica*, 52:681–700, 1984b.

Greene, W. H. *Econometric Analysis*, Sixth Edition. Prentice Hall, New York, 2007.

Gunsolley, J. C., Getchell, C., & Chinchilli, V. M. Small sample characteristics of generalized estimating equations. *Commun Stat – Simul C*, 24: 869–878, 1995. doi:10.1080/03610919508813280

Hall, A. The information matrix test for the linear model. *Rev Econom Stud*, 54:257–263, 1987.

Hall, A. On the calulation of the information matrix test in the normal linear regression model. *Econom Lett*, 29:31–35, 1989. doi:10.1016/0165-1765(89)90169-9

Hall, A. R. Some aspects of generalized method of moment estimation. In Maddala, G. S., Rao, C. R., & Vinod, H. D., editors, *Handbook of Statistics*, Vol. 11, pp. 393–417. Elsevier, Amsterdam, 1993.

Hall, A. R. *Generalized Method of Moments*. Oxford University Press, New York, 2005.

Hall, D. B. On GEE-based regression estimates under first moment misspecification. *Commun Stat – Theor M*, 28:1021–1042, 1999. doi:10.1080/03610929908832341

Hall, D. B., & Severini, T. A. Extended generalized estimating equations for clustered data. *J Am Stat Assoc*, 93:1365–1375, 1998.

Hansen, B. E., & West, K. D. Generalized method of moments and macroeconomics. *J Bus Econ Stat*, 20:460–469, 2002. doi:10.1198/073500102288618603

Hansen, L. P. Large sample properties of generalized method of moments estimators. *Econometrica*, 50:1029–1054, 1982.

Hardin, J. W., & Hilbe, J. M. *Generalized Linear Models and Extensions*, Second Edition. Stata Press, College Station, 2007.

Hausman, J. A. Specification tests in econometrics. *Econometrica*, 46:1251–1271, 1978.

Heagerty, P. J., & Zeger, S. L. Marginal regression models for clustered ordinal measurements. *J Am Stat Assoc*, 91:1024–1036, 1996.

Hin, L. Y., & Wang, Y. G. Working-correlation-structure identification in generalized estimating equations. *Stat Med*, 28:642–658, 2009. doi:10.1002/sim.3489

Hinkley, D. Jackknifing in unbalanced situations. *Technometrics*, 19:285–292, 1977.

Holly, A. A remark on Hausman's specification test. *Econometrica*, 50:749–759, 1982.

Holly, A., Monfort, A., & Rockinger, M. Fourth order pseudo maximum likelihood methods. Research Paper Series 09–23, Swiss Finance Institute, 2008.

Horowitz, J. L. Bootstrap-based critical values for the information matrix test. *J Economet*, 61:395–411, 1994. doi:10.1016/0304-4076(94)90092-2

Horton, N. J., Bebchuk, J. D., Jones, C. L., Lipsitz, S. R., Catalano, P. J., Zahner, G. E., & Fitzmaurice, G. M. Goodness-of-fit for GEE: An example with mental health service utilization. *Stat Med*, 18:213–222, 1999. doi:10.1002/(SICI)1097-0258(19990130)18:2¡213::AID-SIM999¿3.0.CO;2-E

Huber, P. J. *The behavior of maximum likelihood estimates under nonstandard conditions*, Vol. 1, pp. 221–233.University of California Press, Berkeley, 1967.

Jagannathan, R., Skoulakis, G., & Wang, Z. Generalized method of moments: Applications in finance. *J Bus Econ Stat*, 20:470–481, 2002. doi:10.1198/073500102288618612

Kauermann, G. A note on multivariate logistic models for contingency tables. *Aust NZ J Stat*, 39:261–276, 1997. doi:10.1111/j.1467-842X.1997.tb00691.x

Kauermann, G., & Carroll, R. J. A note on the efficiency of sandwich covariance matrix estimation. *J Am Stat Assoc*, 96:1387–1396, 2001.

Kendall, M., & Stuart, A. *The Advanced Theory of Statistics*, Vol. 1, Third Edition. Charles Griffin, London, 1969.

Kuk, A. Y. C. Permutation invariance of alternating logistic regression for multivariate binary data. *Biometrika*, 91:758–761, 2004. doi:10.1093/biomet/91.3.758

Kullback, S., & Leibler, R. On information and sufficiency. *Ann Math Stat*, 22:79–86, 1951.

Küsters, U. *Hierarchische Mittelwert- und Kovarianzstrukturmodelle mit nichtmetrischen endogenen Variablen*. Physica, Heidelberg, 1987.

Lancaster, T. The covariance matrix of the information matrix test. *Econometrica*, 52:1051–1054, 1984.

Lancaster, T. A note on bootstraps and robustness. Centre for microdata methods and practice working paper cwp04/06, Institute for Fiscal Studies, University of Lancaster, 2003.

Laroque, G., & Salanie, B. Estimation of multi-market fix-price models: An application of pseudo maximum likelihood methods. *Econometrica*, 57: 831–860, 1989.

Lee, H. S., Paik, M. C., & Lee, J. H. Genotype-adjusted familial correlation analysis using three generalized estimating equations. *Stat Med*, 27:5471–5483, 2008. doi:10.1002/sim.3344

Lee, K., Joo, Y., Song, J. J., & Harper, D. W. Analysis of zero-inflated clustered count data: A marginalized model approach. *Comput Stat Data Anal*, 55:824–837, 2011. doi:10.1016/j.csda.2010.07.005

Lehmann, E. L., & Casella, G. *Theory of Point Estimation*, Second Edition. Springer, New York, 1998.

Li, K.-C., & Duan, N. Regression analysis under link violation. *Ann Stat*, 17: 1009–1052, 1989.

Liang, K. Y., & Zeger, S. L. Longitudinal data analysis using generalized linear models. *Biometrika*, 73:13–22, 1986. doi:10.1093/biomet/73.1.13

Liang, K. Y., Zeger, S. L., & Qaqish, B. Multivariate regression analysis for categorical data. *J R Stat Soc B*, 54:3–40, 1992.

Lin, D. Y., Wei, L. J., & Ying, Z. Model-checking techniques based on cumulative residuals. *Biometrics*, 58:1–12, 2002. doi:10.1111/j.0006-341X.2002.00001.x

Lipsitz, S. R., Dear, K. B., & Zhao, L. Jackknife estimators of variance for parameter estimates from estimating equations with applications to clustered survival data. *Biometrics*, 50:842–846, 1994a.

Lipsitz, S. R., Kim, K., & Zhao, L. Analysis of repeated categorical data using generalized estimating equations. *Stat Med*, 13:1149–1163, 1994b. doi:10.1002/sim.4780131106

Lipsitz, S. R., Laird, N. M., & Harrington, D. P. Generalized estimating equations for correlated binary data: Using the odds ratio as a measure of association. *Biometrika*, 78:153–160, 1991. doi:10.1093/biomet/78.1.153

Lu, B., Preisser, J. S., Qaqish, B. F., Suchindran, C., Bangdiwala, S. I., & Wolfson, M. A comparison of two bias-corrected covariance estimators for generalized estimating equations. *Biometrics*, 63:935–941, 2007. doi:10.1111/j.1541-0420.2007.00764.x

Magnus, J. R. The asymptotic variance of the pseudo maximum likelihood estimator. *Econ Theory*, 23:1022–1032, 2007. doi:10.1017/S0266466607070417

Mancl, L. A., & DeRouen, T. A. A covariance estimator for GEE with improved small-sample properties. *Biometrics*, 57:126–134, 2001. doi:10.1111/j.0006-341X.2001.00126.x

Mattner, L. Complete order statistics in parametric models. *Ann Stat*, 24: 1265–1282, 1996.

Mátyás, L., editor. *Generalized Method of Moments Estimation*. Cambridge University Press, New York, 1999.

McCullagh, P. Discussion of "Multivariate regression analysis for categorical data." *J R Stat Soc B*, 54:33–34, 1992.

McCullagh, P., & Nelder, J. *Generalized Linear Models*, Second Edition. Chapman and Hall, London, 1989.

Miller, M., Davis, C., & Landis, J. The analysis of longitudinal polytomous data: Generalized estimating equations and connections with weighted least squares. *Biometrics*, 49:1033–1044, 1993.

Molefe, A. C., & Hosmane, B. Test for link misspecification in dependent binary regression using generalized estimating equations. *J Stat Comput Sim*, 77:95–107, 2007. doi:10.1080/10629360600565079

Molenberghs, G. Generalized estimating equations: Notes on the choice of the working correlation matrix. *Methods Inf Med*, 49:419–420, 2010.

Morel, J. G., Bokossa, M. C., & Neerchal, N. K. Small sample correction for the variance of GEE estimators. *Biom J*, 45:395–409, 2003. doi:10.1002/bimj.200390021

Ogaki, M. Generalized method of moments: Econometric applications. In Maddala, G. S., Rao, C. R., & Vinod, H. D., editors, *Handbook of Statistics*, Vol. 11, pp. 455–486. Elsevier, Amsterdam, 1993.

Orme, C. The small-sample performance of the information-matrix test. *J Economet*, 46:309–331, 1990. doi:10.1016/0304-4076(90)90012-I

Paik, M. C. Repeated measurement analysis for nonnormal data in small samples. *Commun Stat – Simul C*, 17:1155–1171, 1988. doi:10.1080/03610918808812718

Pan, W. On the robust variance estimator in generalised estimating equations. *Biometrika*, 88:901–906, 2001. doi:10.1093/biomet/88.3.901

Pan, W., Louis, T. A., & Connett, J. E. A note on marginal linear regression with correlated response data. *Am Stat*, 54:191–195, 2002.

Pan, W., & Wall, M. M. Small-sample adjustments in using the sandwich variance estimator in generalized estimating equations. *Stat Med*, 21:1429–1441, 2002. doi:10.1002/sim.1142

Park, C., & Weisberg, S. Fisher consistency of GEE models under link misspecification. *Comput Stat Data Anal*, 27:229–235, 1998. doi:10.1016/S0167-9473(98)00004-8

Pepe, M. S., & Anderson, G. L. A cautionary note on inference for marginal regression models with longitudinal data and general correlated response data. *Commun Stat – Simul C*, 23:939–951, 1994. doi:10.1080/03610919408813210

Pregibon, D. Goodness of link tests for generalized linear models. *J Appl Stat*, 29:15–24, 1980.

Preisser, J. S., & Qaqish, B. F. Deletion diagnostics for generalized estimating equations. *Biometrics*, 83:551–562, 1996.

Prentice, R. L. Correlated binary regression with covariates specific to each binary observation. *Biometrics*, 44:1033–1048, 1988.

Prentice, R. L., & Zhao, L. P. Estimating equations for parameters in means and covariances of multivariate discrete and continuous responses. *Biometrics*, 47:825–839, 1991.

Qaqish, B., & Liang, K. Y. Marginal models for correlated binary responses with multiple classes and multiple levels of nesting. *Biometrics*, 48:939–950, 1992.

Rao, C. R. *Linear Statistical Inference and Its Applications*, Second Edition. Wiley, New York, 1973.

Rohatgi, V. K., & Saleh, A. K. M. *An Introduction to Probability and Statistics*, Second Edition. Wiley, New York, 2001.

Rotnitzky, A., & Jewell, N. Hypothesis testing of regression parameters in semiparametric generalized linear models for cluster correlated data. *Biometrika*, 77:485–497, 1990. doi:10.1093/biomet/77.3.485

Royall, R. M. Model robust confidence intervals using maximum likelihood estimation. *Int Stat Rev*, 54:221–226, 1986.

Sabo, R. T., & Chaganty, N. R. What can go wrong when ignoring correlation bounds in the use of generalized estimating equations. *Stat Med*, 29:2501–2507, 2010. doi:10.1002/sim.4013

Schildcrout, J. S., & Heagerty, P. J. Regression analysis of longitudinal binary data with time-dependent environmental covariates: Bias and efficiency. *Biostatistics*, 6:633–652, 2005. doi:10.1093/biostatistics/kxi033

Schmoor, C., & Schumacher, M. Effects of covariate omission and categorization when analysing randomized trials with the Cox model. *Stat Med*, 16:225–237, 1997. doi:10.1002/(SICI)1097-0258(19970215)16:3¡225::AID-SIM482¿3.0.CO;2-C

Sharples, K., & Breslow, N. Regression analysis of correlated binary data: Some small sample results for estimating equations. *J Stat Comput Sim*, 42:1–20, 1992. doi:10.1080/00949659208811406

Shults, J. Discussion of "Generalized estimating equations: Notes on the choice of the working correlation matrix" – continued. *Methods Inf Med*, 50:96–99, 2011.

Shults, J., Sun, W., Tu, X., Kim, H., Amsterdam, J., Hilbe, J. M., & Ten-Have, T. A comparison of several approaches for choosing between working correlation structures in generalized estimating equation analysis of longitudinal binary data. *Stat Med*, 28:2338–2355, 2009. doi:10.1002/sim.3622

Snell, E. A scaling procedure for ordered categorical data. *Biometrics*, 20: 592–607, 1964.

Stock, J. H., Wright, J. H., & Yogo, M. A survey of weak instruments and weak identification in generalized method of moments. *J Bus Econ Stat*, 20:518–529, 2002. doi:10.1198/073500102288618658

Stomberg, C., & White, H. Bootstrapping the information matrix test. Working paper 2000–04, University of California at San Diego, Department of Economics, 2000.

Stram, D., Wei, L., & Ware, J. Analysis of repeated ordered categorical outcomes with possibly missing observations and time-dependent covariates. *J Am Stat Assoc*, 83:631–637, 1988.

Sutradhar, B. C., & Kumar, P. On the efficiency of extended generalized estimating equation approaches. *Stat Probab Lett*, 55:53–61, 2001. doi:10.1016/S0167-7152(01)00127-4

Venezuela, M. K., Botter, D. A., & Sandoval, M. C. Diagnostic techniques in generalized estimating equations. *J Stat Comput Sim*, 77:879–888, 2007. doi:10.1080/10629360600780488

Vens, M., & Ziegler, A. Generalized estimating equations and regression diagnostics for longitudinal controlled clinical trials: A case study. *Comput Stat Data Anal*, in press, 2011. doi:10.1016/j.csda.2011.04.010

Wang, M., & Long, Q. Modified robust variance estimator for generalized estimating equations with improved small-sample performance. *Stat Med*, in press, 2011. doi:10.1002/sim.4150

Wedderburn, R. W. M. Quasi-likelihood functions, generalized linear models, and the Gauss-Newton method. *Biometrika*, 61:439–447, 1974. doi:10.1093/biomet/61.3.439

Wei, L., & Stram, D. Analysing repeated measurements with possibly missing observations by modelling marginal distributions. *Stat Med*, 7:139–148, 1988. doi:10.1002/sim.4780070115

White, H. A heteroskedasticity-consistent covariance matrix estimator and a direct test for heteroskedasticity. *Econometrica*, 48:817–838, 1980.

White, H. Consequences and detection of misspecified nonlinear regression models. *J Am Stat Assoc*, 76:419–433, 1981.

White, H. Maximum likelihood estimation of misspecified models. *Econometrica*, 50:1–25, 1982.

White, H. *Estimation, Inference and Specification Analysis*. Cambridge University Press, Cambridge, 1994.

Yan, J., & Fine, J. Estimating equations for association structures. *Stat Med*, 23:859–874, 2004. doi:10.1002/sim.1650

Zeger, S. L., & Liang, K. Y. Longitudinal data analysis for discrete and continuous outcomes. *Biometrics*, 42:121–130, 1986.

Zeger, S., Liang, K., & Self, S. The analysis of binary longitudinal data with time-independent covariates. *Biometrika*, 72:31–38, 1985. doi:10.1093/biomet/72.1.31

Zhang, B. An information matrix test for logistic regression models based on case-control data. *Biometrika*, 88:921–932, 2001. doi:10.1093/biomet/88.4.921

Zhu, H., Ibrahim, J. G., & Shi, X. Diagnostic measures for generalized linear models with missing covariates. *Scand Stat Theory Appl*, 36:686–712, 2009. doi:10.1111/j.1467-9469.2009.00644.x

Ziegler, A. Generalized estimating equations. In Ahrens, W., & Pigeot, I., editors, *Handbook of Epidemiology*, Second Edition. In press. Springer, Heidelberg, 2012.

Ziegler, A. The different parameterizations of the GEE1 and the GEE2. In Seeber, G. U. H., Francis, B. J., Hatzinger, R., & Steckel-Berger, G., editors, *Statistical Modelling Proceedings of the 10th International Workshop on Statistical Modelling*, Vol. 104 of *Lecture Notes in Statistics*, pp. 315–324. Springer, Innsbruck, 1995.

Ziegler, A., & Arminger, G. Parameter estimation and regression diagnostics using generalized estimating equations. In Faulbaum, F., & Bandilla, W., editors, *SoftStat '95. Advances in Statistical Software 5*, pp. 229–237. Lucius & Lucius, Heidelberg, 1996.

Ziegler, A., Kastner, C., & Blettner, M. The generalised estimating equations: An annotated bibliography. *Biom J*, 40:115–139, 1998. doi:10.1002/(SICI)1521-4036(199806)40:2¡115::AID-BIMJ115¿3.0.CO;2-6

Ziegler, A., Kastner, C., & Chang-Claude, J. Analysis of pregnancy and other factors on detection of human papilloma virus (HPV) infection using weighted estimating equations for follow-up data. *Stat Med*, 22:2217–2233, 2003. doi:10.1002/sim.1409

Ziegler, A., Schäfer, H., & Hebebrand, J. Risch's lambda values for human obesity estimated from segregation analysis. *Int J Obes Relat Metab Disord*, 21:952–953, 1997. doi:10.1038/sj.ijo.0800496

Ziegler, A., & Vens, M. Generalized estimating equations: Notes on the choice of the working correlation matrix. *Methods Inf Med*, 49:421–425; discussion 426–432, 2010. doi:10.3414/ME10-01-0026

Index

Printed by Books on Demand, Germany